# Events as a Strategic Marketing Tool

# Events as a Strategic Marketing Tool

*Editor*

**Leo Rask**

scitus academics

**Events as a Strategic Marketing Tool**
Edited by **Leo Rask**

ISBN: 978-1-68117-423-5
Library of Congress Control Number: 2016937578


SCITUS Academics LLC,
www.scitusacademics.com
Box No. 4766, 616 Corporate Way,
Suite 2, Valley Cottage,
NY 10989



**Notice**



Printed in United States of America on Acid Free Paper

# Preface

Have you ever been tempted by a credit card company giving away free blankets or t-shirts at a football game in exchange for a credit card application? Have you been approached by a skimpily-dressed young woman offering samples of fruity alcohol at a local bar? Each of these unique experiences offers a glimpse at what twenty-first century event marketing can be. Experiences, encounters and events have come to play an ever-growing role in marketing as their impact is more fully understood. As a profession, however, event marketing is still a relatively new field, with most available literature covering practical project management rather than the strategy behind including events in a marketing campaign. The events industry now includes events of all sizes from the Olympics down to business breakfast meetings. Many industries, charitable organizations, and interest groups hold events in order to market themselves, build business relationships, raisc money or celebrate achievement. Events are considered one of the strategic marketing and communication tools by companies of all sizes. From product launches to press conferences, companies create promotional events to help them communicate with clients and potential clients. A number of elements such as music, live entertainment or even the particular venue may be used to influence the tone and atmosphere of an event. Event marketing strategies leave a lasting, brand-focused impression of fun by grabbing the attention of a group of people who are gathered together. If executed successfully, event marketing will provide each of them with an experience that will resonate in their minds. The book Event as a Strategic Marketing Tool describes how events can be used as a strategic tool in marketing practices. The book addresses the development of the experience economy, events, and event marketing. Subsequently, the book covers the various areas of marketing within which experiences play a role, such as branding, relationship marketing and city marketing.

# Table of Contents

# CHAPTER 1

# Culture and City Branding: Mega-Events and Iconic Buildings as Fragile Means to Brand the City

**Müge Riza**

Eastern Mediterranean University, Famagusta, North Cyprus

## ABSTRACT

In the last years many cities investigated diverse approaches to attract more audience and stimulate their economy. One of the key terms in this context is city branding. City branding strategies are used as a tool to race with other cities by displaying appealing images of the city and its assets. In this context, culture is as a marketable strength and new opportunity for cities to increasingly gain popularity. This paper deals with the utilization of cultural as a branding device. It discusses the role of culture in city branding and compares the use of recent popular tools, namely large- scale events, festivals and iconic architecture as a means for branding purpose. Finally, the paper argues that iconic buildings and mega-events using permanent structures in contrast to smaller events and festivals are unsustainable means to promote the culture of the city, as most of the time they are undermining the existing local culture in favour of a global taste, broadcasting similar images and activities.

## KEYWORDS

City Branding, Culture, Iconic Architecture, Mega-Events

## 1. INTRODUCTION

Countries and increasingly cities are challenged with globalization and its effects on their international reputation. The fear of economic drawbacks has resulted in a strong competition between cities and nations [1] . Citizens and visitors seem to be enthusiastic about new images, extraordinary buildings and exciting places to live and visit. Additionally, investors chose locations for their business with appealing environment and cultural and social facilities. As a consequence, local governments have started to consider city branding as part of their urban management scheme to attract a broader market [2] [3]. As argued by Lang, a city has to act as a "business company" selling and promoting its goods. However, branding is far more than creating and selling appealing images. It is about the creation of authentic metaphors. Literature reveals that especially culture has developed as a strong "good", which plays a substantial role in city branding [4] [5]. As accurately mentioned by Kavaratzis, "Culture provides consumable experiences; culture acts as a source for economic activity; culture attracts the creative class; culture attracts tourists..." [6]. The research questions are as follows: how culture is used in city branding? Which cultural is going to be branded? Is iconic architecture representing local (material) culture?

The aim of this paper is to develop answers to these questions by exploring the role of material culture in city branding. The article deals with the relationship between city branding and culture. Different approaches on how city branding is utilized through the use of culture will be introduced with a focus on events and iconic structures. Finally, large-scale events and iconic architecture as a controversial tool in city branding are discussed. The paper claims that iconic buildings and mega events using iconic structures are not representing authentic cultural values and leading to similar city images or activities. In this sense this paper tries to evoke question marks about the uncritical usage of these strategies for branding purposes and their sustainability in a long term.

## 2. CITY, CULTURE AND CITY BRANDING

The city has always been a symbol for culture and cultural activity as it is providing more than the basic needs of existence such as shelter and food. Even for the ancient Greeks the city was a synonym for civilized life [7]. As defined by Wells, culture of the city are "those things that make the city distinct from other human settlements there are things in cities that, in general, it is difficult to imagine having in a place outside of the city". Furthermore he argues, that cities themselves might be characterized as cultural artefacts as well [8]. In cities, as mentioned by Zukin, culture is present through the built environment in form of palaces, theatres, museums and temples, and also through parks, marketplaces and memorials [9]. Hence, most cities own a variety of cultural assets that are possible features to utilize for branding purposes. In this sense not only heritage buildings and history are targeted, but other cultural assets are considered as well to develop a strategy to promote the city. As summarized accurately by Ashworth, there are mainly three instruments used by city authorities to increase the worldwide attention of the city: event hall marketing, personality association and flagship and signature district [10]. In the recent agenda of cities, especially two of these strategies are used effectively as branding devices: large-scale/mega events such as European Capital of Culture or the Olympic Games, and iconic architecture [1]. As argued also by Kong, cities with global ambitions have realized "the need to accumulate cultural capital, for which one means is to create new urban spaces, in particular, new cultural urban spaces (e.g. grand theatres, museums, libraries)" [11] . No doubt, in the current worldwide trend of globalisation, cities are in need to get attention as they are bound to and influenced by the global economy and development. However, it is worth to critically examine the usage of these kinds of approaches as a branding tool as there is a danger of reproducing similar images and spectacles around the world without considering the authenticity of the local culture. Therefore, in the upcoming part small scale events will be compared with large-scale events and iconic architecture under their contribution to express and integrate local culture.

## 3. SMALL EVENTS VERSUS MEGA-EVENTS AND ICONIC ARCHITECTURE

In general, events could be characterized through their exclusivity as well as through their temporary presence. Events are designed mainly for a short period in which the city is transformed into a "cultural stage". Especially smaller scale events are attractive as they create a positive image of the city and give opportunities to experience the city as a participant. Additionally, if they are offered continually during a specific period of the year, they might be considered as a tool to gain attention and recognition. The below shown example is a street festival in one of the most popular high streets in Munich, Germany, the Leopoldstrasse (Figure 1). The event is held regularly two times each year and gained popularity for tourists as well as for local visitors. One advantage of these kinds of events is that they are adaptable to the needs of the visitors and are able to handle with changes in the cultural agenda. As they are not manifested through "buildings", they are only temporary transforming or influencing the existing setting—in terms of spatial, visual as well as social conditions (Figure 2).

Large-scale events such as The European Cultural Capital or the Olympic Games have similar roots as small- scale events. They are planned as temporary actions, with the aim to host visitors and participants for a limited period during the year. Indeed, cities could gain positive reputation of such events. For most European Cities the nomination as a "European Cultural Capital" is a prestigious position. To be a cultural capital means to get financial support for cultural activities, development of image and international audience and recognition [5] . In a wider perspective, it appears as if there are reasonable arguments for large events as they provide a variety of benefits for the host city, economic as well as cultural. Problematic are those cases, where the promotion of the cultural event is done at the expanse of the local cultural environment [5]. Unfortunately, in many cases those kinds of events are in contrast to smaller festivals accompanied by several "fixed" means to promote the festivals and events. Zhang and Zhao, in their research of the Olympic Games in Beijing, were exploring the branding of Beijing through the Olympic Games in 2008. According to them Beijing's "identities were branded symbolically through the logo, the theme slogans, the mascots and the like, and materially through construction of landmark buildings and infrastructure". In order to host the athletes the National Stadium was constructed in form of an iconic architectural design, while traditional neighbourhoods were removed [12]. Hence, the construction of the iconic building destroyed existing texture and local culture. In the case of Beijing this approach was disastrous for the residents of the area and generated damage to cultural continuity and identity. As argued above, events do not bear significant harm as long as they are temporary and not presented through architectural manifestation. Instead buildings, ones constructed, will be present in the built environment, as a consistent part of it. The same argument is valid for iconic structures that are not constructed as part of a cultural mega-event. Often city authorities assume that impressive architecture by default is increasing attractiveness and reputation of the city. In fact, this could be the case as long the iconic building is actually unique and identifiable. Unfortunately, cities are seeking for extraordinary architecture, but getting into the trap of exposing similar city images (Figure 3 and Figure 4). Fitting examples for this case are the Guggenheim Museum in Bilbao and the Walt Disney Concert Hall in Los Angeles, both designed by Frank O'Ghery. Although, the buildings were established in different locations and for different purposes, the similarities in terms of form and used material are apparent. As a result the context to the surrounding and to the culture is disregarded, as it is not possible to distinguish neither their location nor their function from their outer appearance. The comparison of the two buildings shows that because of their similarities the context to the urban setting is replaced by the context to the architect. Even more, Bandarin argues that architects are "looking to the wrong direction: inwards themselves, not outwards to the urban contexts in which they are building" [13].

Accordingly, in most cases contemporary iconic architecture is not representing or referring to local culture and authenticity. As argued by several researchers, even though the world is getting closer and more mobile, as an effect of globalization local identity is becoming an important issue for cities [5] [14] . Evans assume, that every city has a "certain cultural character", which makes it irreplaceable and different from other places [15]. Additionally, as argued by Lui, places are carrying meanings, history and local specifications, while especially the "local", and its place-based character, makes it unique [16]. Unfortunately, locality is currently undervalued in many city branding practices and city authorities should pay more attention to the cities exclusiveness rather than to global images. When using iconic structures for branding purposes, it should be noted that a brand itself is a symbol that identifies and ads value. As many cities posses similar characteristics, a city branding strategy have to deal with the specific values of the city to distinguish it [12]. The sole repetition of similar images will lead to loss of (existing) identity and decreases finally the brand value [17]. Therefore any successful city branding has to consider its own local values and cultural peculiarities to establish a sustainable and authentic city brand image. As argued by Evans the "commercial and competitive" city branding is able to create irregular and unexpected outcomes [15] . Therefore it is uncertain how the city will be affected in long term through the use of mega cultural projects including iconic buildings, as they tend to "reinforce a homogenous culture" [15] . More than this, Lui argues that places are not just a site for the installation of world-class facilities and "signature architecture". A place is required to develop its own sense of place, a quality of permanence [16] . Iconic buildings can go anywhere in the world. On the other hand, it should be considered that the "sense of place" is mainly visible in the core of historic cities where the texture is still old, whereas in the new areas of the cities the "sense of place" does not occur in the same presence [18] . Therefore it is important to distinguish in which part of the city the iconic architecture is going to be assembled.

**Figure 1**. Street festival on a main street in Munich, Germany (pictures: personal archive).

## 4. CONCLUSION

It is outlined in this paper that city branding through culture is a complex and controversial approach. Especially the use of materialized culture in form of mega events and iconic structures to promote the culture of a city may be accompanied by unpredictable risks. An iconic building, successful in getting attention for a specific purpose in a specific place, will be repeated in other locations regardless of the existing background. In such cases the culture of the local vicinity is not considered or even ignored. Contextual qualities, cultural and local issues are put into the background in favour of global representations. As a consequence all over the world we are facing the trend towards unified city images and similar public spaces. As branding is about the creation of authentic and distinct images, construction of similar images would end up in most cases in short lasting success. Ignoring the local culture in advance of appealing "global images" would finally end up in the creation of same city images, which at the end would not articulate any local culture. That in turn would affect cities in a long term negatively as they could lose their authenticity, which is essential for successful city branding. City governors should be aware of the qualities and values of their city and choose a branding device that fits particularly to their case. Otherwise city images would be interchangeable and the place would lose attractiveness for tourists and visitors. Likewise locals can feel themselves alien with the building or events that are not representing their culture and habits.

**Figure 2**. Temporary stage pictures: personal archive.

**Figure 3**. Guggenheim museum in Bilbao (picture: personal archive).

**Figure 4**. Walt Disney Music Hall in Los Angeles (picture: personal archive).

## REFERENCES


1. Kavaratzis, M. (2005) Branding the City through Culture and Entertainement. The AESOP 2005 Conference, Vienna, 13-18 July 2005.

2. Salo, L. (2012) Building a Creative City Brand through an International Mega-Event. Case: World Design Capital Helsinki 2012. Master Thesis, Turku School of Economics, Turku.

3. Eshuis, J. and Edelenbos, J. (2009) Branding in Urban Regeneration. Journal of Urban Regeneration and Renewal, 2, 272-282.

4. Hornskov, S.B. (2007) On the Management of Authenticity: Culture in the Place Branding of Oresund. Place Branding and Public Diplomacy, 3, 317-331.

5. Kunzmann, K.R. (2004) Culture, Creativity and Spatial Planning. Town Planning Review, 75, 383-404.

6. (Kavaratzis, M. and Ashworth, G. (2015) Hijacking Culture: The Disconnection between Place Culture and Place Brands. Town Planning Review, 86, 155-176.

7. Short, J.R. (1996) The Urban Order. An Introduction to Cities, Culture and Power. Blackwell Publishers, Cambridge.

8. Wells, K. (2007) The Material and Visual Cultures of Cities. Space and Culture, 10, 136-144.

9. Zukin, S. (2004) Dialogue on Urban Cultures: Globalistaion and Culture in an Urbanizing World. UN-Habitat World Urban Forum, Barcelona, 13-17 September 2004.

10. Ashworth, G.J. (2009) The Instruments of Place Branding: How Is It Done? European Spatial Research and Policy, 16, 9-21.

11. Kong, L. (2007) Cultural Icons and Urban Development in Asia: Economic Imperative, National Identity, and Global City Status. Political Geography, 26, 383-404.

12. Zhang, L. and Zhao, S.X. (2009) City Branding and the Olympic Effect: A Case Study of Beijing. Cities, 25, 245-254.

13. Bandarin, F. (2006) Urban Conservation in an Age of Globalization.http://www.ihbc.org.uk/context_archive/97/bandarin/edinburgh.html

14. Zukin, S. (1995) The Culture of Cities. Blackwell Publishing, Oxford.

15. Evans, G. (2003) Hard Branding the Cultural City: From Prado to Prada. International Journal of Urban and Regional Research, 27, 417-440.

16. Lui, T. (2008) City-Branding without Content. Hong Kong's Aborted West Kowloon Mega-Project, 1998-2006. IDPR, International Development Planning Review, 30, 215-226.

17. Riza, M., Doratli, N. and Fasli, M. (2012) City Branding and Identity. Procedia—Social and Behavioral Sciences, 35, 293-300.

18. Cresswell, T. (2004) Place: A Short Introduction. Blackwell Publishing, Oxford.

# CHAPTER 2

# Test a Queue Detection System for Special Events in Nevada

**Hualiang (Harry) Teng[1], Xuecai Xu[2], Venkatesan Muthukumar[3], A. Reed Gibby[4]**

[1]Department of Civil and Environmental Engineering and Construction, University of Nevada, Las Vegas, USA
[2]School of Civil Engineering and Mechanics, Huazhong University of Science and Technology, Wuhan, China
[3]Department of Electrical and Computer Engineering, University of Nevada, Las Vegas, USA
[4]Department of Civil and Environmental Engineering, University of Nevada, Reno, USA

## ABSTRACT

Queue detection systems have been used in traffic management in work zones and have also been recommended for traffic control for special events like sports and conventions. However, they have not been tested in the field. This paper presents the results of tests for a queue detection system at two special events in Las Vegas, Nevada.

The system consisted of two vision detectors, radio frequency communications and one changeable message sign. Two aspects of the system were evaluated: the effectiveness of the system in reducing speeds and the cost and effectiveness of its deployment. In the tests, traffic data such as queue length and vehicle operating speeds were collected and analyzed to see whether motorists respond congestion related message on the changeable message sign when they did not perceive the congestion. By this approach, the motorists' true responses to the system were identified. The results indicated that motorists did make positive responses to the messages provided by the system. However, it was found that the system may not be cost-effective because significant costs would be incurred in purchasing, installation and maintenance of the system. Recommendations were provided to utilize existing message signs and detectors to provide the same information to motorists as a queue detection system does.

## KEYWORDS

ITS Technology; Special Events

## 1. INTRODUCTION

It is very common in peak periods that queues back up from some freeways to connected arterials, or vice versus at some locations in Nevada so that queues interface with through lane traffic. One phenomenon is that there are many entertainment facilities in Las Vegas that organize major events (e.g., conventions) which tend to cause short term surges in traffic, and the resulting queues often become hazardous. Note that these situations pose even greater dangers during non-peak periods. It is because drivers in these non-peak periods may not be from the local areas and thus may not be as familiar to roadways as the drivers in peak periods (during which most of the drivers are commuters). To prevent crashes in this situation, an effective approach is to install queue detectors at the potential locations and alert the upstream motorists of the queues using technologies such as changeable message signs. With the messages displayed on a changeable message sign, motorists can take due precautions. By doing so, the number of crashes in this situation can be reduced.

Basically, a queue detection system can be applied to the following four conditions:

- Congestion related to recurring traffic conditions [1-4]
- Congestion related to work zones [5]
- Congestion related to incidents [6], and
- Congestion related to special events such as sporting events and conventions.

The time and place for recurring congestion, work zones and special events can be known in advance and thus the queue detection system can be installed based on the evaluation of anticipated traffic congestion conditions. However, it is difficult to anticipate in advance the time and place for incidents in advance, and thus incident response vehicles typically carry portable queue detection systems (primarily consisting of a CMS) to incident scenes to alert motorists of incidents downstream. So far, most of the applications of queue detection systems have been for recurring congestion, work zone and incidents. Applications of the queue detections to special events have been recommended in a guidebook for managing traffic for special events, and the queue detection system has not be tested for a special event. This study was the first attempt to test the queue detection system in selected locations for special events.

A typical queue detection system consists of detectors, radio frequency communication and variable message signs. Detectors are to collect data of traffic variables such as volume, speed and presence from which queue length can be estimated. If a queue is detected, the messages on variable message signs installed upstream will be adjusted correspondingly to alerting the motorists of the queue downstream. The radio frequency communication transmits the information of queue detection from the detectors to the display controllers attached to changeable message signs. Detectors can be different types including tube, loop, radar and camera.

The evaluation of the queue detection system in this study was not on the system reliability or maintenance issue. It was focused on whether motorists made response to the message displayed on the changeable message sign. There have been studies on evaluating the effectiveness of message sign. It has been not clear whether the responses by motorists were to the perceived congestions or the displayed message. This study particularly made this distinction in the evaluation.

The deployments of the queue detection system have shown significant effectiveness in reducing the number of crashes. For instance, INDOT installed a special warning system at an exit on I-65 in 1999. The number of crashes was decreased from twenty-eight occurred between May and December 1998 to sixteen during the same period in 1999. Like many technologies in the Intelligent Transportation Systems, queue detection systems for special events involve the installation of the system in an area associated with special events. The deployment of such a system was not as easy as deploying a speed trailer in a work zone. The installation and maintenance of the system in the field may be very timeconsuming. Particularly, the time period during which the queue detection system can be effective for a typical deployment is usually short and thus may not be able to justify its applications for special events. Thus, cost and benefit analysis was conducted for this system in this study. Note that there are traffic control plans developed for every special event. However, these plans only cover the area close to the places where the activities of special events occur. The impacted traffic usually located far beyond the area covered by typical traffic control plans.

To achieve the objective of the study, a literature review was performed primarily on the application of queue detection systems in work zones because queue detection has been mostly applied in work zones. Based on the literature review, it was identified that queue detection system has been tested in several states that showed positive results in reducing speeds. Different from most of the queue detection system where microwave non-intrusive detectors were used in queue detection system, this study developed a system that used vision detection technologies for monitoring traffic conditions at identified location. The developed system was tested in Sam Boyd Stadium and in the Thomas & Mack Center for a football game and Bull Rider competition, respectively. Vision technology was also used to collect traffic data for observing the performance of the system in term of the motorists' compliance to the messages displayed on changeable message signs. The data derived from using the vision detection technology were analyzed based on the measurements of effectiveness: speed, queue length and volume. It was found that vehicle speeds can be reduced significantly even when motorists did not see the actual queues. Benefits and cost analyses were performed for both test sites. It was found that deploying such a queue detection system may not be cost-effective due to the high cost of the system that cannot be balanced by the saving for avoiding the occurrence of some crashes at the identified locations for a special event. Thus, the study recommended to use technologies such as changeable message signs and traffic sensors that have been in place to reduce system costs. That way, the cost of deploying the queue detection system can be decreased, for which the development of queue detection system can be justified.

In the following part of the paper, the second section covers the literature review. The third section details the methodology for conducting the tests. The fourth and fifth sections present the tests at two locations, respectively. In Section 6, the cost and benefit for adopting the queue detection system for special events are covered. Conclusions and recommendations were provided in the last section.

## 2. LITERATURE REVIEW

Queue detection system is perceived effective in improving safety at recurrently congested location, work zone, incidents and special event. There have been many applications of the queue detection system to work zones. Thus, the literature review in this section focuses on the literature on work zone related application only.

The queue detection system deployed in work zones is also called automatic work zone information system (AWIS). It can include different system components, dependent upon whether the system intends to disseminate congestion information to the vicinity of deployed changeable message signs or to a larger area beyond the work zone area. The typical components of the systems are traffic data collection, communications, and dissemination system. The data collection systems are located in work zones where road and traffic conditions are different from regular conditions. The collected data are distributed to motorist in different ways, depending on the spatial location of the motorists the system intends to serve.

So far, the following systems manufactured by different companies have been tested in previous studies: TIPS (Traffic Information and Prediction System), ADAPTIR (Automated Data Acquisition and

Processing of Traffic Information in Real-time), CHIPS (Computerized Highway Information Processing System, and Intelli Zone. Each of them has been tested in more than one state. In addition to these major systems, there were other systems that were developed for research or tested less extensively. They were also reviewed in this study.

TIPS was developed by Dr. Prahlad D. Pant of the University of Cincinnati. It was jointly evaluated in[7] and [8]. Basically, the system was designed to collect speed data and disseminate travel time information to motorists traveling immediately upstream of work zone. Through changeable message signs, every motorist passing the signs can be informed of travel time through work zone downstream. Relatively short range communications were employed for passing information between where the information was collected and where the information was accessed. TIPS have been tested in other studies [9-11]. Horowitz et al. [9] evaluated the TIPS from the perspective of its impact on traffic diversion. It was found that the diversion was not significant. Pigman and Agent[10] evaluated the TIPS in the following aspects: 1) performance and reliability, 2) travel time estimation, 3) diversion, 4) crash data, and 5) driver opinion.

ADAPTIR was developed by Scientex Corporation. McCoy and Pesti [12] evaluated the ADAPTIR in Nebraska. The system collected speed data using RTMS which were mounted on the top of portable CMS. The collected speed data were processed to determine the messages to be displayed on each CMS, which were located at different places before a work zone. The messages displayed on CMSs were speed advisory in nature. The system in McCoy and Pesti [12] was evaluated primarily on whether it can reduce vehicle speed. It was found that motorists reduced their speed during congestion when they were aware of the presence of work zone. It was not known whether the reduction of speed was caused by the presence of congestion or the messages on portable CMSs.

In [13], the ADAPTIR was evaluated in Arkansas. The system used RTMS to collect speed data, which were then transmitted to CMS for display. In addition, the speed data and relevant information was transmitted to two highway advisor radio stations for broadcasting. Furthermore, the speed and traffic information were sent to selected group of staff via pager. In their study, Tudor et al. (2003) focused on operation and safety issues.

CHIPS was a product from the ASTI Transportation Systems. The system tested in [14] used Remote Traffic Microwave Sensors (RTMS) to detect whether there was congestion at several places before a work zone. The congestion information was transmitted to portable CMSs on sites or a system at a central location. That way motorists both on site and off site can receive the messages and made response correspondingly. Trailer mounted CCTV cameras were also installed at some places, and live videos were transmitted to a remote central location from which motorists off the work zone can obtain traffic information in the work zone. In the test, traffic information was also sent to motorists through email and paging. In [14], the effectiveness of the CHIP was evaluated based on the travel time of traffic running through work zone, diversion rate at off ramp before the tested work zone, and some surrogate measures of crashes.

The CHIPS evaluated in [13] had similar system components as the ADAPTIR jointly tested in their study. Instead of using landline to transmit information to HAR in ADAPTIR, it used cell phone. In addition, it also included an email service to disseminate the traffic information (traffic condition, delay, and diversion advisories) to a selected group of staff. Like their evaluation for the ADAPTIR, Tudor et al. [13] evaluated CHIP with the focus on operation and safety issues.

IntelliZone was an AWIS that was manufactured by Quixote Transportation Safety [15]. In this system, magnetic detectors were installed in each lane to measure traffic variables such as speed and volume. A mobile command center located on site received the collected traffic data through wireless communications and decided on the messages (primarily related to speed) to be shown on the portable CMS. The system was evaluated for its impact on speed and speed variance. This system was also evaluated in [16]. Instead of using magnetic detectors, this system integrated microwave detectors for measuring speed in their study.

The D-25 Speed Advisory Sign System from MPH Industries was tested in [17]. It is a system that was similar to the one tested in [12], except the speed sign on the speed trailer. When the downstream speed differentials was less than 15 mph, two strobes, each on one side of the speed trailer would flash, and the measured speeds would also be displayed. When the downstream speed differential was greater than 15 mph, the slowest speed measured or the speed limit would be displayed with flashing. The purpose of the study was to evaluate the effectiveness of the speed trailer in reducing speed and speed differential upstream a work zone. Note that the effect of the sign with flashing was not separately evaluated. It was evaluated as part of a whole operation scenario.

Research Build System: There were systems that were developed by research teams for their evaluation. An example was the system developed in [18]. Video detection technology was used to collect traffic data such as speed and vehicle classification. These traffic data were transmitted to a controller where they were analyzed to determine the messages to display on CMSs that were deployed in advance of a work zone. Additionally, a website was developed for broadcasting traffic information in work zone to a wider area. In this study, Pesti et al. [18] focused their effort on the diversion issue, with an extensive review on previous studies related to evaluating the diversion impact of the AWIS.

It can be seen from the literature review that there have been several manufacturers providing AWIS: TIPS, ADAPTIR, CHIPS, and IntelliZone. AWIS consists of some basic system components: traffic data collection, data transmission, controller, message presentation. A simple AWIS consists of detectors, wireless communications, a controller, and CMSs, all on site. The purposes of such an AWIS could be speed advisory, travel time and delay information provision, and incident alerts. The popular traffic data collection technologies are portable non-intrusive detectors such as RTMS. Magnetic sensors which were used in a study are intrusive in nature, and may not be easy to install in a work zone. Video detection may be a choice, but may suffer from the data reliability problem. Speed was the most basic information that can be produced from the system. Other information such as queue, congestion, travel time, travel delay, and incidents have to involve transforming the speed data collected at several places within or upstream a work zone. The transformation of the data may bring in the reliability issue to the derived information. In most of cases, all the detectors were placed upstream a work zone. The number of detectors and portable changeable message signs could be varying, depending upon the accuracy of the information the system intends to provide. The major aspects that the system was evaluated were speed control, traffic diversion, crashes, operation reliability, and information accuracy.

A more complicated AWIS involved disseminating traffic information to a large area beyond work zones. Such information would be helpful in trip planning such as route choice or departure time choice. The popular ways to disseminate the information were highway advisory radio, web site, fax, email, and paging. The population of the motorists being able to access the information varies from some chosen staff to travelers in a large area. The evaluation of such a system is very challenging. The system evaluated in this study was a simple system that only allows dissemination of queue related information in the vicinity of the CMS in the system.

# 3. METHODOLOGY

## 3.1. System Development

The system developed in this study included a vision detection system that was installed at places where queues usually formed. This vision detection system consisted of two cameras, each covering a number of traveling lanes, depending up the locations where the tests were conducted. The image data from the cameras were transmitted via Radio Frequency to Autoscope that was collocated with a changeable message sign upstream of queues that were supposed to occur. The image data were then processed and analyzed to monitor the traffic at the camera locations. A computer program was developed to make decision on whether the traffic was in congested or non-congested conditions. This program was also able to determine the messages to be presented on the message board. A connection was also made between the program and the message board, which allowed the recommended messages to be displayed appropriately. The program was written such that the messages on the message board can be changed manually.

## 3.2. Choice of Test Sites

In this study, four facilities have been considered as test sites: Las Vegas Speedway, Las Vegas Convention Center, Thomas & Mack, and Sam Boyd. The basic criteria used to choose a facility to conduct the test are:

1) There are queues formed during a special event that are frequently occur and be stabilized at relatively a predictable location on a roadway segment for a noticeable time period;

2) There are utilities such as light poles, sign cranes and traffic control poles near the location that allow to mount the components of the queue detection system such as cameras and RF antennas.

It is required to have a queue that fluctuates at a fixed location because the queue detection system tested in this study works only at a predetermined location. If the queue progresses quickly in a vicinity of a location such as a ramp or an intersection, the queue detection system can exhibit its function only for a short period of time, for which the effectiveness of the system for a sustainable period cannot be evaluated.

Several visits to each candidate facilities were made. Before the visits to a facility, meeting with the traffic control personnel of the facilities were held for information on the location where queues frequently appear. The background information on the traffic control plan for special events held in a facility can also be obtained which would help determine the situation of congestion in special events. Visits to the identified sites were made later on to observe the progression of queues and the availability of utilities to install the system components. Usually, a sketch of the identified location was developed. This sketch was used to determine the location of system components. In particularly, the following considerations were given: 1) Location of the CMS, 2) Location of detectors, and 3) Communications between CMS and the detectors. Finally, it was Thomas & Mack Center and Sam Boyd Stadium were chosen as the test sites, one on the freeway and the other on street arterial.

## 3.3. Field Data Collection and Data Analysis

Vision detection technology was used to collect traffic data for system evaluation. Cameras were installed on light poles before and after the changeable message sign. The videos recorded were processed in-house for outputting the traffic data. The evaluation of the queue system for special events was focused on testing whether the system can truly be effective in reducing speeds. To evaluate whether motorists slowed down for the messages on the message board, the time period showing the message "Congestion Downstream" was split into at least two parts - one with queue that cannot be seen by the drivers, and the second with queue that appear in eyes of the drivers. The differences between traffic variables such as queue length during these time phases were used to demonstrate the effectiveness of the queue detection systems on slowing down traffic. Whether the differences between the traffic variables were significant statistically was tested using hypothesis testing method.

## 3.4. Cost and Benefit Analysis

The objective of the benefit and cost analysis was to determine the characteristics (duration and length of impact area) of the special events that can result in more benefits than the cost incurred for installing such a queue detection system. Thus, the benefits and costs were calculated for individual special events.

The benefit that can be realized due to the reduction of speeds of vehicles at special events and thus the corresponding reduction of accidents can be estimated based on costs that would be saved for not having these accidents happening. According to [19], "a 1 km/h change in speed can expect to result in a 3% change in the number of crashes". Thus, if the total change in speed is expressed as $\Delta_v$ (mph), the change in the number of crashes can be estimated as:

$$\Delta_{crash} = \frac{\Delta_v}{0.6213} \times 0.03 \times N_{crash} \tag{1}$$

where $N_{crash}$ denotes the total number of crashes estimated to happen in a special event. In this study, $\Delta_v$ was the speed reduction caused by deploying the queue detection system. In addition, only two categories of crashes were considered: fatal and non-fatal. If needed, more categories of crashes can be adopted. With the total number of crashes estimated for these two categories of crashes, the number of crashes that can be avoided due to the speed reduction can be calculated using the equation below:

$$\Delta_{fatal} = \frac{\Delta_v}{0.6213} 0.03 \times N_{fatal} \tag{2}$$

$$\Delta_{non\text{-}fatal} = \frac{\Delta_v}{0.6213} 0.03 \times N_{non\text{-}fatal} \tag{3}$$

Since crash rate is known (at least on a national level) for different categories of crashes, the total number of crashes in these categories can be derived for a road segment on which traffic travel for a special event as follows:

$$N_{fatal} = r_{fatal} VMT \tag{4}$$

$$N_{non\text{-}fatal} = r_{non\text{-}fatal} VMT \tag{5}$$

$$N_{\text{crash}} = N_{\text{fatal}} + N_{\text{non-fatal}} \tag{6}$$

where $r_{\text{fatal}}$ and $r_{\text{non-fatal}}$ are crash rates for fatal and non-fatal crashes, respectively; $r_{\text{fatal}}$ and $r_{\text{non-fatal}}$ denotes the crash rates for these two categories of crashes, and VMT represents the vehicle miles traveled by the vehicles through the location where congestion occur during special events.

For a road segment carrying special event traffic, the VMT can be estimated as:

$$VMT = AADT \times L \times D \tag{7}$$

where $L$ denotes the distance over which vehicles keep their reduced speeds. The term $D$ represents the duration during which speeds were reduced. It is usually short (about half hour in this study) and can be represented as a percentage of 24 hours.

If the unit costs for these two categories of crashes are available, the total cost saving (denoted as B for benefit) due to avoiding the occurrence of these crashes can be estimated as:

$$B = c_{\text{fatal}}\Delta_{\text{fatal}} + c_{\text{non-fatal}}\Delta_{\text{non-fatal}} \tag{8}$$

where $c_{\text{fatal}}$ and $c_{\text{non-fatal}}$ represent the unit costs for these two categories of crashes, respectively. From [20], it can be found that the unit costs for the crashes with severity level of K, A, B, Code, O, are $3,654,299, $181,276, $46,643, $22,201, and $2116, respectively. The unit cost for the fatality is $3,654,299, while that for the non-fatality crashes was derived as the average of the other four unit costs, which gives $63,059.

The total cost is the sum of the following four categories of costs: capital (C), installation, operation, and maintenance. The capital cost C can be annualized as:

$$C_a = C\frac{i(1+i)^n}{(1+i)^n - 1} \tag{9}$$

where i represents the interest rate and n is the life cycle. In the calculation in this study, the interest rate was assumed to be 8% for all the system components, and the life cycles of different system components were assumed the same as 10 years. Because a special event would be held for a week at most, the annualized capital cost was converted to a daily based value:

$$C_d = C_a/365 \tag{10}$$

Then, the capital cost for a special event last several days can be derived as:

$$C_p = C_d \times D \tag{11}$$

The system components included in the calculation were: CMS, cameras, video detection processor, and radio frequency. One CMS and two RF were counted, while two video detection processors and two cameras were included. The unit costs for CMS, camera, video detection processor and RF were $50,000, $400, $4000, and $300, respectively. The present value of the capital cost was calculated as $59,100. Since their life cycles were assumed the same and be equal to 10 years, the annualized cost can be calculated as $8808 at 8% interest rate. This annualized cost can be used to derive the daily based capital cost $C_d$.

Installation cost was the cost to set up the system for a special event $C_{inst}$. The installation included mounting and dismounting cameras and RF, and system integration in the field. Mounting the cameras and RF would take about three hours with the involvement of one ITS related technician and one maintenance crew consisting of one supervisor and three workers. Dismounting these cameras and RF would take approximately another one hour. Integrating the system making sure the system works as expected would take two more hours in the field. The annual salaries for both the ITS technician and the supervisor of the maintenance crew were calculated at $60,000, and the salary for the three maintenance workers was estimated at $40,000. The fringe benefit rate was assumed to be 30%. Then, the cost for setting up and dismounting the system $C_{inst}$ was calculated at $1503.

It was also assumed that the ITS technician should be available for operating the system during about half of the duration of a special event. The cost for this part of work comprised the operation cost $C_{opr} = h \times D$, where $h$ is the technician's hourly rate.

The maintenance cost was estimated as the half of the capital cost, which can be written as:

$$M_p = 0.5 \times C_d \tag{12}$$

The total cost for having such a system in a special event can be derived as:

$$T_p = C_d \times D + C_{inst} + h \times D + 0.5 \times C_d \times D \tag{13}$$

It can be seen from the equation that only the installation cost is not a function of the impact time period of special events. Given the benefit (B) and cost $(T_p)$ estimated above, the benefit and cost ratio can be derived as $B/T_p$.

## 4. TEST ON SAM BOYD STADIUM

Figure 1 displays several access routes to Sam Boyd Silver Bowl. The formal address for this facility posted on its website was: 7000 E Russell Rd, Las Vegas, Nevada 89122-8338. Thus, this facility was mostly accessed from Tropicana Avenue and Russell Road. Since Russell Road exit was suggested as the best candidate for testing the technology, the system was set up over there. Another condition making this site feasible for testing the system was that there were facilities on which the cameras and antennas can be mounted.

It was indicated by the Manager of Special Events in the Thomas & Mack Center that congestion usually backed up to freeway I-515 from the off ramp at Russell Road during events such as football games. Traffic controls for such event were usually set up at locations close to Sam Boyd Stadium, not to the area around the freeway. Thus, motorists either attending the special events or passing the off ramp always ran into the congestion with no warning.

Figure 2 displays the configuration of the system at the site. Cameras 2 and 3 were part of the vision detection system. Two Autoscope image processors were connected to these two cameras (one for each of Cameras 2 and 3, respectively) to collect traffic data such as speed, occupancy, and volume. Figure 3 shows utility pole where Camera 2 was mounted. One antenna was also installed jointly with the camera and the image processor at each of these two locations. These antennas were used to transmit the traffic data to a RF at the location where a CMS was deployed.Figure 3 also presents the plastic pole carrying the RF that was attached to the CMS. The RF was connected to a computer as a controller. A program was written in the controller to determine the appropriate messages to be displayed based on analyzing the traffic data received from the RF transmissions. The space between the RF at CMS and the RFs at the other two locations were clear and an eye sight between them was available for communications. There are two more cameras at the site: Camera 4 at CMS and Camera 1 at the location where Camera 2 was mounted. They were used to record traffic, which was then processed to derive the performance measures for evaluating the performance of the queue detection system.

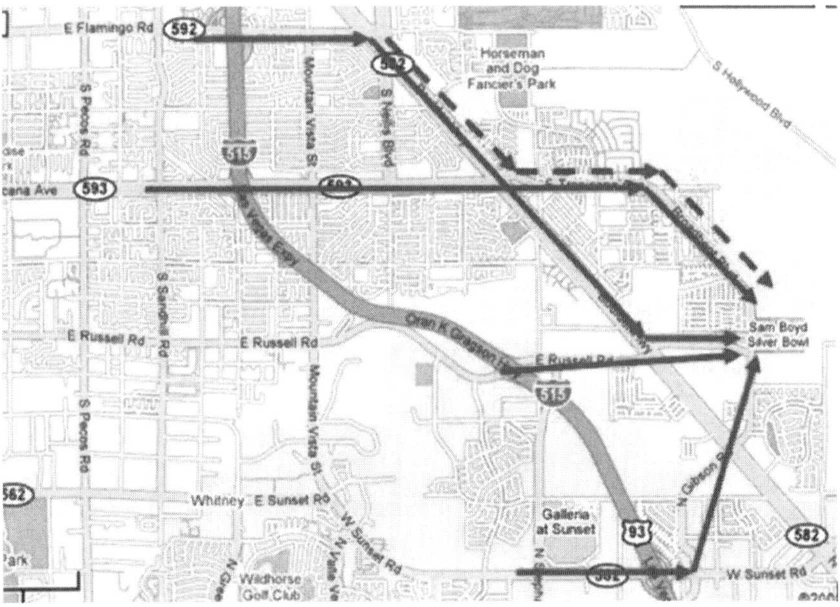


**Figure 1**. Access to Sam Boyd Silver Bowl (Google map).

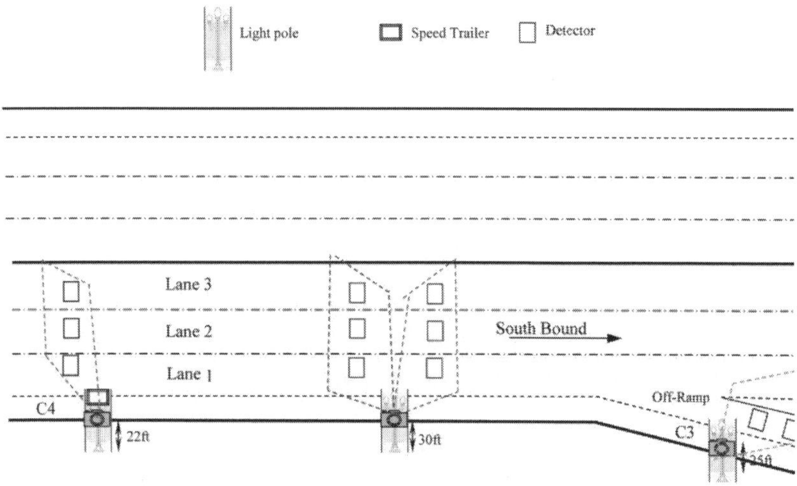


**Figure 2**. System layouts and evaluation equipment at Russell Rd. off-ramp.

**Figure 3**. Components of the queue detection systems in the field. (a) Plastic pole attached to a utility pole carried Cameras 1 and 2 and a RF. (b) Camera mounted on a plastic pole attached to the CMS trailer. (c) CMS displaying "congestion right lane".

## 4.1 Field Tests—September 15, 2007

The football game on September 15, 2007 started on 6:30 pm. The system was set up at 4:45 pm. From 4:45 pm to 5:10 pm, no congestion was observed on the off ramp. During this time period, no message was displayed on the CMS. From 5:10 pm to 5:30 pm, the message "Drive Safely" was displayed because no congestion was observed on the off ramp. Starting from 5:30 pm, congestion was observed on the off ramp, but did not back up to the mainline freeway. Thus, the message "Congestion Right Lane" was displayed since then. At 5:40 pm, it was seen that the congestion spilled to mainline freeway and motorists can see the congestion from upstream of the CMS. The congestion recessed back to the off ramp at 6:10 pm. The congestion stayed on the off ramp until 6:30 pm. From 5:30 pm to 6:30 pm, the message "Congestion Right Lane" was displayed for the entire time period. The congestion was first not perceivable from the mainline, then perceivable, later not be seen again.

Figure 4 presents the patterns of speeds over the three lanes at the location where Cameras 1 and 4 were installed. From the upper chart in Figure 4 it can be seen that the speeds when the message was "Drive Safely" was smaller than those when the CMS showed no message. When congestion started showing on the off ramp, the message turned to "Congestion Right Lane". The speeds were reduced further down. After the congestion was presented on the mainline freeway, the speeds were reduced additionally, which is the speeds shown at the bottom in the upper chart in Figure 4. When the congestion shrank back to the off ramp, the speeds went up to a level similar to the case when there was congestion at the beginning or the case when there was no congestion.

From the chart (for Camera 1) in the lower part of Figure 4 it can be seen that the similar pattern of speed changes can be observed over these different time periods with different levels of congestion and messages shown on the CMS, One difference was that the speeds in these time periods at this location were lower than those at the upstream Camera 4. These results from Cameras 1 and 4 have been tested using the hypothesis test method. The test results in Table 1 confirm the results observed from Figure 4.

Note: Ramp Congestion 1 and Ramp Congestion 2 represent congestions happened on the off ramp before and after the congestion spilled to freeway, respectively. "<" indicates that the speed in the time period shown on the left most column is the same as the time period indicated on the corresponding heading. This definition is the same for ">" and "=".

## 5. TEST AT THE SITE FOR THE THOMAS & MACK CENTER

Visits were made to two special events held at Thomas & Mack for observing the traffic congestion on Tropicana Avenue (see Figure 5). Queue was seen fluctuating around the Pacific Terminal, very close to the Koval Ln. Intersection. This observation was briefed to the Thomas & Mack and was confirmed. It was also confirmed at the site that there were tall light poles that Clark County allowed to install cameras and antennas.

The site at the Thomas & Mack Center was chosen on Tropicana Avenue from Las Vegas Blvd to Paradise Road (see Figure 5). During special events such as Professional Bull Rider, queues were observed to form at the intersection of Paradise Road and Tropicana Avenue and backed up to the intersection at Koval Ln.

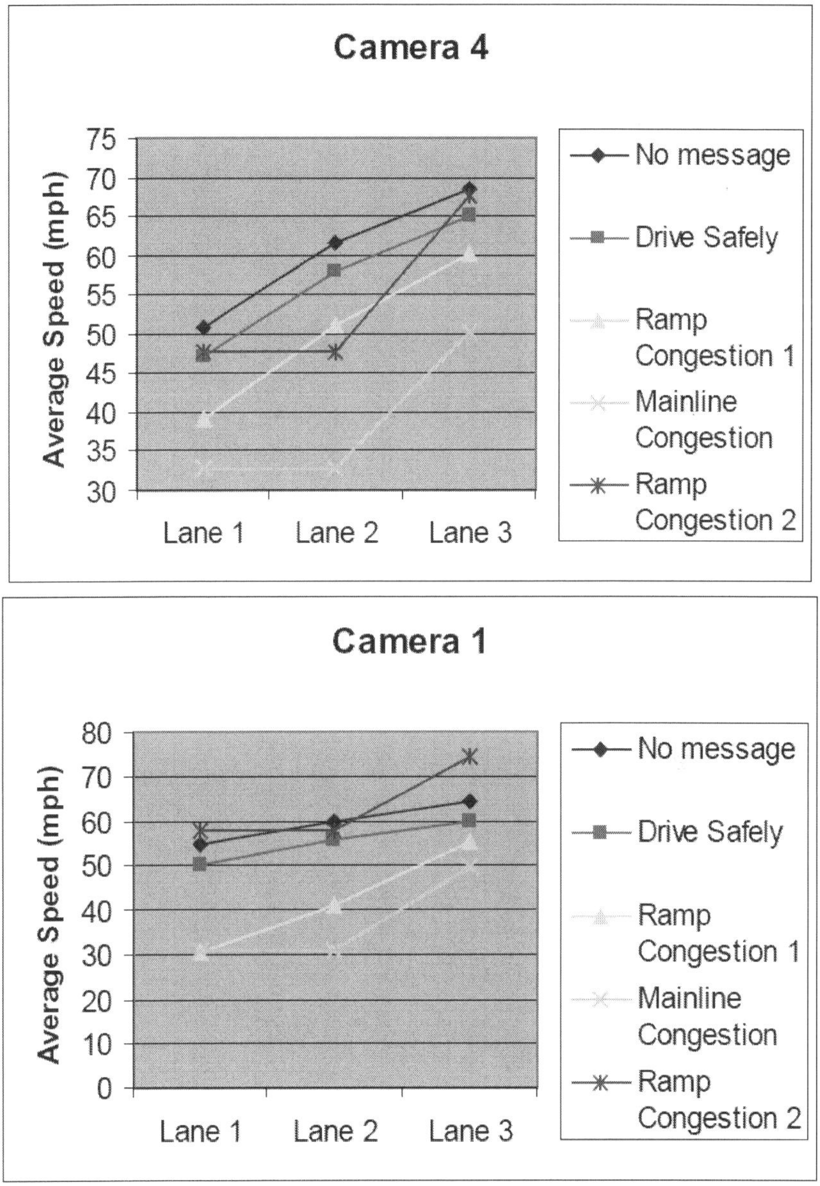


**Figure 4**. Speed data from Cameras 1 and 4 on December 15, 2007.

**Table 1**. Hypothesis test results for the speeds of traffic responding to different messages on September 15, 2007.

| | | | | |
|---|---|---|---|---|
| | | | CAM 4 | |
| | | Lane 1 | Lane 2 | Lane 3 |
| No Message | | | No Message | |
| Drive Safely | | < | < | < |
| | | | Mainline Congestion | |
| Congestion Right Lane | Ramp Congestion 1 | > | > | = |
| | Ramp Congestion 2 | > | > | > |
| | | | CAM 1 | |
| | | Lane 1 | Lane 2 | Lane 3 |
| No Message | | | No Message | |
| Drive Safely | | < | < | < |
| | | | Mainline Congestion | |
| Congestion Right Lane | Ramp Congestion 1 | > | > | = |
| | Ramp Congestion 2 | > | > | = |

To detect the queue backed up from Paradise Rd. and monitor the progression of the queue to Koval Ln, two cameras (Cameras 1 and 2 in Figure 6) were mounted on street light poles. The CMS was deployed on the median shoulder at the intersection of Island Way, facing the parking garage of MGM Casino and Hotel on Tropicana Avenue (see Figure 7). From Figure 5, it can be seen that the road segment on Tropicana Avenue from Las Vegas Blvd to Paradise Road was straight, between which there was no object blocking the eye sight for the communications between the RFs that were installed at Cameras 1 and 2 and the CMS.

To collect data for evaluating the performance of the queue detection system, Cameras 3 and 4 (see Figure 8) were mounted on two light poles, one before and one after Island Way. In the test on I-515, one camera was mounted on a plastic pole that was attached to the trailer that carried the CMS. In the test at this site, however, a camera was mounted on street light which provided the stability for the camera to collect stable images of videos.

## 5.1 Field Test on November 3, 2007

The tests at this site were conducted on November 3, 2007 for the special event Professional Bull Rider. That day, the event started on 6:00 pm. The system was ready by 3:20 pm for collecting data before traffic started entering the game. From 3:20 pm to 3:30 pm, no message was displayed on the CMS. By the time 3:30 pm, the message changed to "Drive Safely". From 4:45 pm to 6:00 pm, the message "Congestion Left Lane" was exhibited. During this one hour and fifteen minutes, congestion conditions were changing dramatically. From 4:45 pm to 5:20 pm, queues fluctuated before the Koval Ln Intersection. From 5:20 pm to 6:00 pm, the queues backed to the location close to where the CMS was located.

The videos collected from Cameras 3 and 4 were processed to derive the measures of effectiveness: queue length and speed, for different time periods, each corresponding to different messages displayed on the CMS and the level of congestion condition. From Figure 5, it can be seen that traffic had to make left turn at Swenson Avenue to enter Thomas and Mack. Thus, queue was formed on the left lane. The message "Congestion Left Lane" was expected to direct traffic not attending the special event to use the right lanes. Thus, the traffic before the intersection at Island Way on Tropicana Avenue may distribute differently over the three lanes comparing with the case with no message displayed. Thus, the queue lengths before the start of the green light on the three lanes on the eastbound approach at the Island Way intersection were manually counted through replaying the recorded videos on computer. In addition, the speeds of vehicles on these three lanes during the green phase of signal were derived using the video image processor included in Autoscope. With the traffic data-queue length and speed, collected, the mobility of traffic and the resultant safety impact were evaluated. From Figure 7, it can be seen that a bus stop was located very close to the area where motorists were expected to make response to the CMS. It was observed in the field that vehicles had to wait behind a bus if there was no gap on the next lane for them to change lane. Vehicles could fully utilize the space occupied by a bus only when buses left the stop. Thus, these two performance measures were collected with the distinction whether there were buses presenting at the scene. According to the messages shown on the CMS, there are distinct three time periods: no message, "Drive Safety", and "Congestion Left Lane". To show whether a visible queue had more impact on motorists' responses to a message, two time periods were distinguished for the time period when the message "Congestion Left Lane" was displayed. The first time period was corresponding to the duration when a queue did not pass the Koval Ln intersection. The second was when the queue reached to the area close the CMS and thus motorists can see the queue clearly.

From Camera 3, the speed and traffic volume data were collected for the green light time periods because this camera is at the downstream location of the Island Way intersection. Queue length data that can be collected from Camera 4 cannot be observed from this camera. Traffic volumes among these three lanes can be more revealing to the impact of the CMS. The speeds measured at this location can be viewed as the continuation from the location at Camera 4.

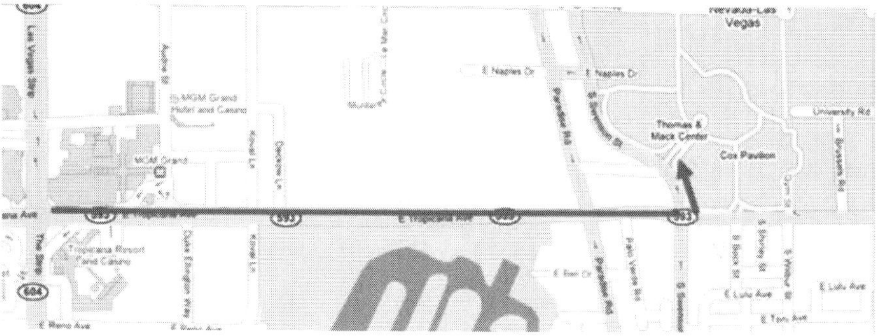


**Figure 5**. Direction of traffic to Thomas & Mack.

**Figure 6**. System layouts in the field on Tropicana Avenue.

Figure 9 presents the queue length collected from Camera 4. The upper chart for no bus at the bus stop indicates that there were longer queues in left lanes and right lanes. The queue lengths during the congestion time period were significantly longer than the other two time periods. The queue lengths during the two time periods corresponding to congestion visible (close) and invisible (far) were very close. A tendency that more traffic shifted to right lanes can be observed when the message showed "Congestion Left Lane", which implies that motorists spread out over the lanes even they did not see the congestion ahead.

**Figure 7**. Locations of a CMS and a bus stop.

From the lower part of Figure 9, which displays the speed data from Camera 4, it can be seen that the speeds on the right lanes were higher than the left lanes, which is consistent with our intuitive. In addition, the speeds during the congestion period were lower than those in non-congestion periods. This observation reflected the impact of congestion on speeds. The speeds during the congestion period when congestion was far from the CMS were lower than those when congestion was close to the CMS. Note that these differences in speeds in these two conditions are on the magnitude of 2 to 3 mph, which is small.

**Figure 8**. Cameras for collecting observational data for system evaluation.

Figure 10 exhibits the speeds and traffic flow measured at Camera 3, downstream of Camera 4. It can be seen from the upper chart that the speeds during the time period when no message was displayed on the CMS were higher than those when there were message displayed. The speeds when congestion was visible to the motorists close to the CMS were lowest. The speeds in the other two time periods were in between of these two cases, which implies that any message, either "Drive Safely" or "Congestion Left Lane", was effective in slowing down vehicles' speeds. It can also be observed from the chart that the speeds when the congestion was not visible at the CMS were higher than that during the "Drive Safely" was shown. This indicates that the message "Congestion Left Lane" was effective in spreading vehicles over the traveling lanes and thus increased vehicles' speeds. From the chart at the bottom ofFigure 10 it can be seen that flows during the time periods when no message was shown and "Congestion Left Lane" was shown for visible congestion were either on the top or at the bottom levels. Those flows for the other two conditions were in between. This observation is an additional indication that support the findings based on the speeds at this location.

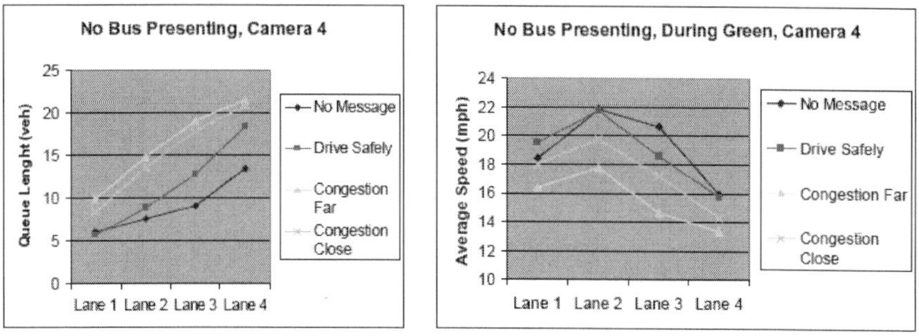


**Figure 9**. Queue length and speeds from Camera 4 on November 3, 2007.

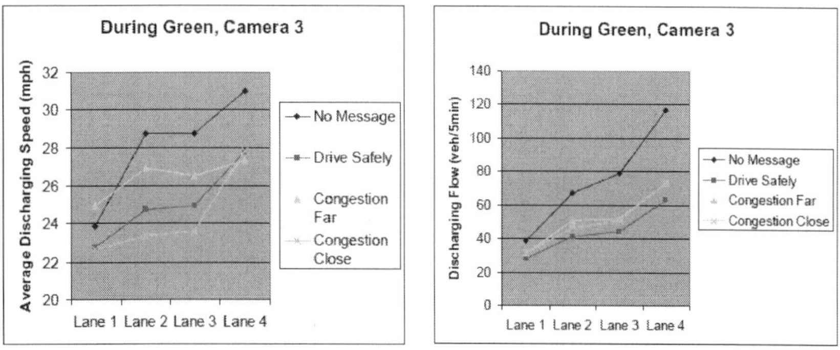


**Figure 10**. Speeds and flows at Camera 3 on November 3, 2007.

# 6. COST AND BENEFIT ANALYSIS

In calculating the benefit for the site at Sam Body, the AADT was taken as 124,667. The speed reduction was 10 mph for each of the three lanes. From Sinha and Labi (2007), it was found that the fatality and non-fatality rates for interstate are 0.56 and 46.56 per 100 million VMT, respectively. It was assumed that the queue detection system demonstrated its unique contribution in reducing speeds only during the time period when there were congestions on the off ramp. Speeds were also reduced during the congestion presented on mainline. However, the speed reduction during this time period when there was no such a queue detection was installed can be viewed as the same as that period when there was such a queue detection system in place. It was assumed that the time period with the unique contribution by the system was 2 hours (the actual duration was 20 minutes). The traffic would sustain the reduced speeds up to 1500 ft. The B/C was calculated as 0.02. Note that the case assumed for the calculation was developed based on the test on I-515. To conduct a sensitive study on the B/C versus the total time period when the queue detection is the most effective, the B/C's was calculated for the special event that lasted up to five days. B/C varies from 0.02 to 0.09. This result indicates that the queue detection system at the test site was not cost-effective.

It can be found that the reason for the system being not cost-effective is the high costs incurred in planning, designing, operating and maintaining the system. To see the condition of a queue detection system with varying level of complexity, the queue detection system was simplified by including only a CMS and two operators in the field. The system works by manually changing the messages on the CMS with the traffic report from the other operators close to the site where congestion usually appears. The B/C was calculated as 0.35, still less than 1.

For the test site at Thomas & Mack, the same method as used for Sam Boyd was adopted. Since this site was on arterial, the crash rates for fatal and non-fatal crashes were assumed to be 1.3 and 124.69, respectively. The range on which vehicles could sustain their lowered speed was 1500 ft. The time period during which the queue detection system can be really effective was two hours. The benefit and cost ratio was calculated as 0.02, far less than 1, making the cost and benefit even. Considering the queue detection system being deployed for 100 times in a year, the benefit and cost ratio is still almost the same as 0.02.

Basically, there are several variables that are critical in determining whether the queue detection system is costeffective: the scale of the impacted area, the traffic volume which influenced the duration during which the queue detection system was effective and the crash rate. Compared to the work zones for which the queue detection system can be applied, the major difference is the duration during which the system is effective. In the calculation above, two-hour was used, which was longer than the actual duration time. It was noted that a work zone could last for days and months. During this extended time period, the traffic running through a work zone reduced their speed. Thus, the VMT used to calculate the number of crashes would be significantly high. Relatively, the VMT used to derive the number of crashes for special events would be small. Thus, the benefit generated due to avoiding the crashes would be small. This makes the benefit and cost ratio to be smaller than one. Based on this analysis, it can be proposed that a good strategy for improving safety in special events is to use the technologies that are in place already to alert the motorists of the emerging congestions.

# 7. CONCLUSIONS AND RECOMMENDATIONS

In this study, the effectiveness of queue detection system in reducing speed was evaluated by focusing on the performance of the system during the time period when motorists can only see the message, but not the congestion. In addition, a cost and benefit study was conducted to evaluate the conditions that such a queue detection system can be beneficial. The following findings can be concluded. First, motorists responded to congestion message by reducing speeds in some traveling lanes. This finding can be found in other studies in evaluating the effectiveness of queue detection system. It seems not beneficial economically to deploy the queue detection system for controlling speed at locations on the approaches to special events. The reason is that the duration of traffic experiencing speed reduction was short which made it difficult to generate enough benefits that would be produced by avoiding occurring crashes at the location approaching to special events.

It was recommended that the technologies existing on the current roadway system be better used to alert traffic queue and congestion. By using the existing technologies, the costs for alerting motorists of upcoming congestion can be reduced significantly. Two important components in a queue detection system are changeable message sign for disseminating traffic information and detectors for monitoring traffic condition. If a changeable message sign exists at a location very close to the location where queue would back up during special events, this message sign can then be used to display congestion related warning message. Less expensive detectors can be deployed at identified location to monitoring traffic. If available, existing detectors can also be used for monitoring traffic condition. Usually, changeable message signs and sensors comprise a significant portion of the total cost of the system. With using the existing changeable message sign and traffic sensors, the total capital cost of the whole system can be reduced substantially.

It should be aware that some parameters used in the calculation may have certain variation. For example, the crash rates used in the calculation may be on the lower side. They may be very high at the location close to special events. There might be other saving caused by diverting traffic to different routes. The task including this in the calculation is challenging and thus not presented in this study.

## ACKNOWLEDGEMENTS

This study cannot be completed without the supports from many agencies and individuals. The research team would like to give their thanks to the following individuals for their assistance during the project: NDOT District 1 and Clark County, David Stetzer of Las Vegas Speedway, Michael Thompson of Traffic Control Service, Inc., Steve Patterson of the Las Vegas Visits and Convention Authority, and Kenny Sasaki, and Xin Li of UNLV.

## REFERENCES

1. S. C. Tignor, et al., "Innovative Traffic Control Technology and Practice in Europe," 1999.http://www.international.fhwa. dot. gov/Pdfs/Innovtce.pdf

2. Indiana Department of Transportation Press Release, "Clark County I-65 Exit 4 Solution Proves Successful," 2000. http://www.state.in.us/dot/motoristinfo/trafficwise/wnew_otherindiana_news_04.html

3. King County Department of Transportation Press Release, "New Traffic Warning System Improves Safety Near Elementary School,"

2002.http://www.metrokc.gov/kcdot/news/2002/nr020207_208th.htm

4. P. B.Wiles, S. A. Cooner, C. H. Walters and E. J. Pultorak, "Advance Warning of Stopped Traffic on Freeways: Current Practices and Field Studies of Queue Propagation Speeds," FHWA/TX-03/4413-1, 2002.

5. FOCUS, Federal Highway Administration, "Queue Length Detector Reduces Risk of Rear-End Accidents in Work Zones," 1998.http://www.tfhrc.gov/focus/archives/Fcs498/048chips.htm

6. MacCarley, "Evaluation of Caltrans District 10 Automated Warning System: Year Two Progress Report," California PATH Research Report No. UCB-ITS-PRR- 99-28, 1999.

7. P. D. Pant, "Implementation and Testing of the Travel Time Prediction System (TIPS)," Final Report Prepared in Cooperation with the Ohio Department of Transportation and The US Department of Transportation, Federal Highway Administration, 2001.

8. H. T. Zwahlen, "Evaluation of a Real-Time Travel Time Prediction System in a Freeway Construction Work Zone," Final Report prepared for the Ohio Department of Transportation, 2001.

9. J. Horowitz, I. Weisser and T. Notbohm, "Diversion from a Rural Work Zone Owing to a Traffic-Responsive Variable Message Signage System," TRB 2003 Annual Meeting CD-ROM, Washington DC, 2003.

10. J. G. Pigman and K. R. Agent. "Evaluation of Traffic Information and Prediction System as Work Zone Traffic Control," Research Report to Kentucky Transportation Research Center, 2004.

11. K. K. Dixon, "Evaluating Speed Reduction Strategies for Highway Work Zones (Smart Work Zones)," Final Report, Prepared for Georgia Department of Transportation Prepared by Georgia Transportation Institute Georgia Institute of Technology, 2005.

12. P. T. McCoy and G. Pesti, "Effectiveness of ConditionResponsive Advisory Speed Messages in Rural Freeway Work Zones," Transportation Research Record, Vol. 1794, 2002, pp. 11-18.

13. L. H. Tudor, A. Meadors and R. Plant II, "Deployment of Smart Work Zone Technology in Arkansas," Transportation Research Record, Vol. 1824, 2003, pp. 3-14.

14. L. Chu, H.-K. Kim, Y. Chung and W. Recker, "Evaluation of Effectiveness of Automated Workzone Information Systems," Transportation Research Record, Vol. 1911, 2005, pp. 73-81.

15. T. King, C. Sun and M. Virkler, "Intellizone, Report to Widwest Smart Work Zone Initiative," FHWA Pool Fund Study, 2003.

16. J. Horowitz and T. Notbohm, "Evaluation of Intellizone: A System for Providing Speed Advisories to Drivers Entering Work Zones," Report to Widwest Smart Work Zone Initiative, FHWA Pool Fund Study, 2003.

17. G. Pesti, "D-25 Speed Advisory System," Report to Widwest Smart Work Zone Initiative, FHWA Pool Fund Study, 2002.

18. G. Pesti, P. T. McCoy, M. D. Meisinger and V. Kannan, "Work Zone Speed Advisory System, Report to Widwest Smart Work Zone Initiative," FHWA Pool Fund Study, 2002.

19. J. Stuster, Z. Coffman and D. Warren, "Synthesis of Safety Research Related to Speed and Speed Management," Publication No. FHWA-RD-98-154, 1998.

20. K. C. Sinha and S. Labi, "Transportation Decision Making: Principles of Project Evaluation and Programming," John Wiley & Sons Inc., Hoboken, 2007.

# CHAPTER 3

# Assessing the Web-Based Destination Marketing Activities: A Relationship Marketing Perspective

**Liza M. Cobos[a], Youcheng Wang[a*] & Fevzi Okumus[a]**

Rosen College of Hospitality Management,
University of Central Florida, Orlando, Florida, USA

## ABSTRACT

This study aims to assess the Web-based destination marketing activities employed by American Convention and Visitors Bureaus (CVBs). Empirical data was collected via a survey from 260 CVBs in the USA. The research results reveal that organizational size, financial resources and management team's technological expertise are the dominating factors affecting the effective implementation of each of the four functions of Web-based marketing activities (i.e., information, communication, transaction, and assurance) as well as the overall effectiveness of these activities. The findings suggest that CVBs should use Web-based marketing activities under the guidance of relationship marketing principles. However, the research findings further imply that this is a challenging process which requires investment of considerable resources and organizational support. This study contributes to the body of knowledge by providing empirical evidence on this relatively under researched area. The research findings will be of interest to destination marketing organizations.

## KEYWORDS

Relationship marketing, destination marketing, tourism marketing, convention and visitor's bureaus, Web-based marketing

## 1. INTRODUCTION

Destination marketing practices are greatly influenced by advances in information technology (IT) due to the fragmented and information intensive nature of destination products (Buhalis, 1988; Wang & Russo, 2007). Developments in IT are implemented by destination marketing organizations (DMOs) such as convention and visitors bureaus (CVBs) to fully utilize their features in promoting their destinations (Gretzel, Yuan, & Fesenmaier, 2000). In other words, integration of IT systems into the organizational structure and marketing systems is an important requirement for DMOs (Wang & Fesenmaier, 2006). In addition, IT can facilitate the relationship building process with customers by providing systems which collect customer information and translate it into benefits for both the organization and the customer (Zineldin, 2000). The information gathered through technology allows DMOs to tailor their products and services for their potential customers (Ahn, Kim, & Han, 2003).

Indeed, the effective use of Web-based marketing activities is pivotal not only for marketing and promoting destinations but also for creating a competitive advantage for them (Buhalis, 2000). The key to successful online destination marketing efforts depends primarily upon the integrative application of destination information provision, communication mechanisms, e-commerce functions, and relationship building (Wang & Russo, 2007). However, an examination of DMOs' Web sites at different levels reveals that their online destination efforts are still dominated by the traditional mass marketing philosophy with a focus of broadcasting information to the general market (Wang & Fesenmaier, 2006). Obviously, this marketing practice has not taken the consumer's unique needs and wants into consideration, which can substantially compromise DMOs' ability to establish long-term relationships with consumers. In an increasingly competitive marketplace, customers face a variety of choices when buying a product or service. Consequently, organizations seek to fulfill their customers' needs and wants while selling their products and services at a profit. When travelers decide to make a leisure or business trip, they have different options to facilitate the information search and buying process for their travel arrangements. The CVB at a destination is one of the important information sources for consumers to use in their decision-making process. Since the goal of CVBs is to promote and attract visitors to the area by marketing the destination and its services, it is important to pay close attention to its marketing practices. Web-based marketing activities can assist CVBs not only in informing and attracting potential customers to their destinations but also in building long term relationships with them. In short, Web-based marketing activities should be designed and implemented by following the relationship marketing perspective.

Taking a relationship marketing perspective and considering the logical progression of customer relationship building, this study proposes that DMOs' use of Web sites as their major destination marketing systems follows an evolution of four stages: (a) information provision; (b) communication; (c) transaction; and (d) assurance. These four stages represent a hierarchical progression of technology sophistication, interactivity and complexity (Hanson, 2000; Sharma, 2002), which in turn has a positive relationship with the value creation process and the overall success of the Web marketing efforts (Ditto & Pille, 1998;Wang & Fesenmaier, 2006). Thus, the purpose of the study is: (a) to evaluate Web-based marketing activities used by DMOs from a relationship marketing perspective; (b) to measure the effectiveness of Web-based marketing activities implemented by DMOs in each of the four areas; and, (c) to evaluate the impact of organizational factors on DMOs' level of implementation of Web-based relationship marketing activities.

## 2. THEORETICAL BACKGROUND

Grönroos (1990) defines relationship marketing (RM) as a "process of identifying and establishing, maintaining, enhancing, and when necessary terminating relationships with customers and other stakeholders, at a profit, so that the objectives of all parties involved are met" (p. 5). Berry (1995) explains that there are three levels of RM, and each has a different impact on an organization's competitive advantage. Level one of RM depends primarily on pricing strategies, such as emphasizing on pricing incentives to secure customer's loyalty. Level two depends primarily on social bonds, which involves the personalization and customization of the relationship. Level three of RM is mostly dependent on providing structural solutions to important customer problems. In addition to financial (level one) and social bonds (level two), Berry proposes that services provided in level three give the organization a foundation for a strong and difficult strategy for competitors to copy, therefore providing the company with a strong competitive advantage.

Relationship marketing allows the service provider to gain more knowledge about the customer and their requirements and needs (Berry, 1995; Grönroos, 1990). Therefore, an increase in the knowledge on the customer needs and wants and constant customer contact allows the service provider to tailor or customize the service to the customer's specifications (Berry, 1995). Under this context, the development of IT has tremendous impact on marketing activities. Innovations in technology provide new ways to obtain, collect and analyze customer data, communicate with customers, and offer them customized solutions (Vesanen & Raulas, 2006). Indeed, developments in IT are the most important factors in creating, developing, and maintaining long term relationships with consumers (Zineldin, 2000).

Research has demonstrated that developments in IT present opportunities for organizations to create new relationships with consumers (Zineldin, 2000). Programs like data mining tools and data warehousing techniques allow firms to identify and analyze consumer needs (Kim, Suh, & Hwang, 2003). These systems can be used to implement and support a RM strategy (Kim et al., 2003). In other words, advances in IT allow organizations to move from segmenting markets by groups to segmenting by individuals (Berry, 1995). IT increases the practical value of RM by allowing the organization to efficiently perform RM tasks such as tracking buying patterns, customizing services and promotions, coordinating and integrating delivery, providing two-way communication, augmenting service offerings, and personalizing service encounters appropriately (Berry, 1995). According to Buhalis (1998), the strategic use of IT has a big impact on the tourism industry in several functions such as communication and improved efficiency of operations. Due to the essential role information plays in the description, promotion, distribution, organization, and delivery of tourism products and services, technology has become a strategic weapon and a main source of sustainable competitive advantage for tourism organizations (Buhalis, 2000). However, research has found that many tourism businesses use the Internet to provide just information rather than for acquiring information from customers, which leads to lost opportunities of utilizing IT as an affordable, feasible and powerful tool for managing customer relationships.

Researchers have tried to model the different levels of implementation of technology applications in Web sites that assist relationship marketing activities. For example, Hanson (2000) describes the three stages of Web site development: publishing sites (Stage 1, only provides information to the customer); database and forms (Stage 2, combines the ability to provide information and the ability to retrieve information in response to customer's request); and, personalization (Stage 3, creates Web sites catering to a specific individual preferences with the main focus of relationship building). Hanson explains the evolution of Web site use as a tool for implementing Web-based marketing activities. Most organizations begin their Internet marketing efforts with a Stage 1 Web site since it is inexpensive and easy to develop. Many companies make the transition to Stage 2 in which interaction and e-commerce activities are provided. However, very few companies achieve the personalization of Stage 3 as it is more complicated to implement since additional information is required from the customer to provide a customized site.

Alternatively, other researchers propose different views of the evolution and use of the Internet. For example,Contractor, Wasserman, and Faust (2006) explain that the Internet adoption occurs in three stages: substitution, enlargement, and reconfiguration. At the substitution stage, technology is used to perform the same organizational activities; however, efficiency is achieved allowing the organization to expand activities. For example, e-mails requesting information are received and answered through the Web site instead of receiving postal mail. At the enlargement stage, organizations get acquainted and comfortable with utilizing technology; this process allows for the organization to expand the use of technology to other functional areas. A good example would be the use of e-mails as a communication tool to promote more frequent communication between the customer and the company. At the reconfiguration stage, new technology is implemented and integrated into the business process of the organization. This means that completely new systems and technologies are implemented in the organization to accomplish new tasks in new ways. For instance, a company uses a new system to manage customer accounts. This stage requires more skills and continuous learning to manage this transition.

Several researchers have proposed models to explain RM and its implementation on the Web. Sharma (2002) states that an organization's Internet presence evolves in five stages: information, communication, transactions, relationships, and e-commerce. The five stages demonstrate the progress and increased complexity of the Internet functions and how they are used to create value for the customer. Sharma (2002) proposes that as an organization's Internet functions evolve through the five stages, they provide greater value to the customer. This evolution is consistent with Hanson's (2000) model of continuum to explain the impact personalization has on RM. On one side of the continuum, there is homogeneity and no personalization in the marketing efforts or product/service offerings. As the organization makes efforts to differentiate and customize product/service offerings, the customer experience progresses towards the relationship building side of the continuum. In other words, personalization becomes a competitive advantage for a company when it is used to form and maintain a relationship (Hanson, 2000).

Kotler, Bowen, and Makens (2003) propose that there are five basic levels of relationships that can be formed with a customer online: basic (the company sells the product but does not follow up in any way); reactive (the company sells the product and encourages the customer to call at any time with questions or problems); accountable (a company representative contacts the customer before and during the service encounter requesting suggestions for improvement); proactive (the salesperson or company representative contacts the customer from time to time with suggestions, improvements or creative suggestions for the future); and partnership (the company works closely with the customer and other customers to discover ways to deliver better value).

A further review of the literature on Internet applications and implementation reveals that organizations adopt Web site functions in various degrees of sophistication to provide different capabilities to the customer (Doolin, Burgess, & Cooper, 2002; Hanson, 2000). In the relationship building process, Web sites are used by organizations for the following purposes: (a) to communicate with the customer and provide them with information to assist information search and decision making process (Buhalis, 1998; Doolin et al., 2002; Lexhagen, 2005; Sharma, 2002;Subramanian, Shaw, & Gardner, 2000); (b) to sell directly to the customer (Lexhagen, 2005; Doolin et al., 2002; Sharma, 2002;Subramanian et al., 2000); and (c) post-transaction communication or customer service/support (Buhalis, 1998;Doolin et al., 2002; Lexhagen, 2005; Sharma, 2002). The potential of the Internet is appealing to many sectors of the tourism industry, which are dedicated to building long term relationships with customers.

The various models discussed previously help shed light on how tourism organizations use technology to conduct RM activities. Unfortunately, the use of Web applications by DMOs has mainly focused on information provision since the main focus of these organizations has been to just provide information to the public.

DMOs have failed to exploit the full benefits of creating a relationship with the customer through their Web marketing activities due to the limited use of the Web. Relationship-building can be achieved by allowing interactive communication between customers and the organization, allowing transactions to be completed, and providing personalization/customization capabilities and customer loyalty or retention programs. DMOs should strive to implement RM functions to create long term relationships with customers, provide a better customer experience and create greater customer satisfaction to build long lasting relationships.

These observations have also been supported by Ritchie and Ritchie's (2002) argument that the deployment of destination Web sites encompasses not only the informational aspects of a destination's products, but also the marketing and communication components. Based on the multi-faceted tasks of CVBs, it is argued in this study that in order to build long term relationships with consumers, a successful destination Web site depends on the integrative application of four components as its major functions: (a) Timely and accurate representation and provision of destination information (information); (b) effective and constant communication with consumers (communication); (c) reliable and seamless electronic transaction deployment (transaction); and (d) effective and lasting relationship building mechanisms (assurance) (see Figure 1). This proposed model of Web-based RM activities for CVBs are consistent not only with the philosophy and logic of RM literature (e.g., Gummesson, 1994), but also with research findings in the area of Internet technology implementation which suggests that organizations implement Web-based technologies in stages following a hierarchical progression of technology sophistication, interactivity and complexity (Contractor et al., 2006; Hanson, 2000; Sharma, 2002).

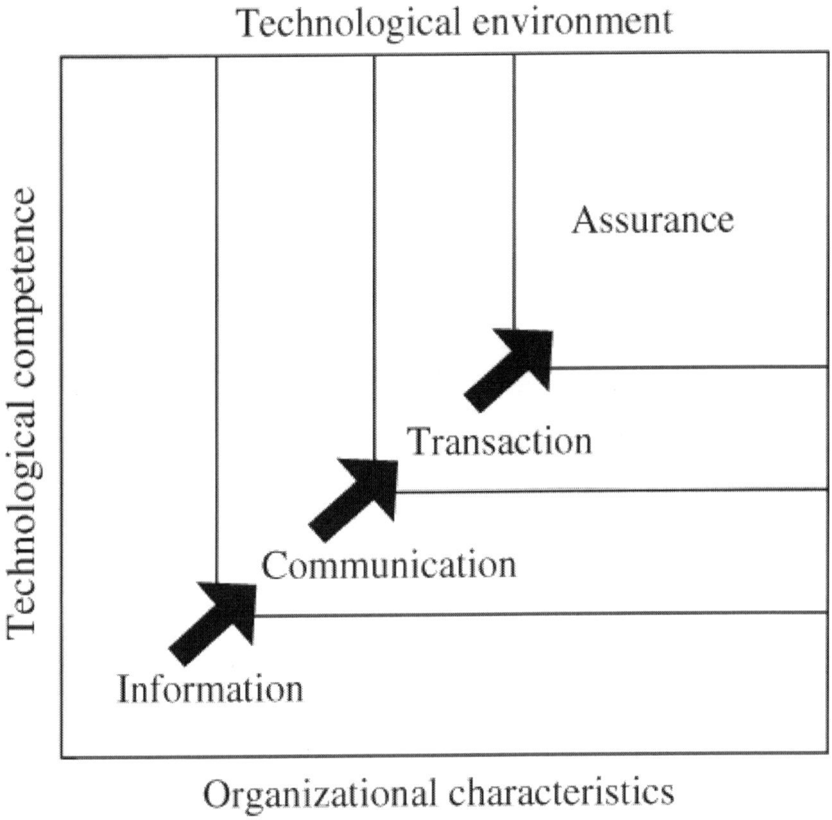


**Figure 1.** Proposed conceptual model of DMOs Web-based relationship marketing implementation and the impact of organizational factors.

The effectiveness of implementing the above four RM functions as well as the overall effectiveness of Web marketing strategy are influenced by the organizational factors such as organizational size, financial resources, technology experience and managerial support. Studies on the innovation adoption, implementation and diffusion field have found that organizational factors both facilitate and inhibit Web marketing strategies. Previous research on innovation adoption and diffusion has focused on both the attributes of the innovation and the characteristics of the organization (Damanpour, 1991;Frambach, 1993; Tornatzky & Klein, 1982). In this context, organizational variables such as size, being receptive to change, competitive environment, strategic direction, management team's characteristics and government regulation have been the focus of investigation in innovation adoption and diffusion research (Davis, Bagozzi & Warshaw, 1989; Damanpour, 1991;Frambach, 1993; Lefebvre & Lefebvre, 1992; Lefebvre, Mason, & Lefebvre, 1997; Rogers, 1995; Thong, 1999;Tornatzky & Fleischer, 1990; Zhu & Kraemer, 2005). Based on an extensive literature review and considering the unique context of this study, the following three groups of organizational factors are deemed important in affecting the effectiveness and sophistication of CVBs' Web-based marketing activities: technological environment, organizational characteristics, and technological competence. The full proposed conceptual model is presented visually in Figure 1.

To summarize, this study proposes that CVBs conduct Web-based RM activities following a hierarchical progression of four functions (i.e., information, communication, transaction, and assurance). The study also argues that the effective implementation of these four Web-based marketing functions as well as the overall effectiveness of Web-based marketing activities are affected by three groups of organizational factors: (a) technological environment (i.e., IT training, financial resources, management involvement, and management support); (b) organizational characteristics (i.e., organizational size and innovativeness); and (c) technological competence (i.e., management technological expertise and employee technological expertise).

# 3. RESEARCH METHODOLOGY

## 3.1 Development of the Research Instrument

As proposed in the conceptual model, Web-based RM activity is composed of four interrelated functions (i.e., information, communication, transaction, and assurance), and each of them contains multiple applications. First, a list of items was identified for each of the four functions based on an extensive literature review and observation of destination marketing organizations' Web sites at different levels. A panel of 10 experts (10 graduate students enrolled in an information technology and tourism class) was then consulted to confirm the appropriateness of the list for each area, resulting in the identification of 12 applications for information function, 5 applications for communication function, 7 applications for transaction function, and 8 applications for assurance function (See Table 2 for a detailed list).

Since one of the aims of this study was to examine the impact of organizational factors on the effective implementation of the four Web-based marketing functions, an effective index was created for each of the four functions. For this purpose, the extent of usage and perceived importance of each of the applications in the four respective functions were evaluated based upon responses to two questions: (a) whether or not (0 = *No*, 1 = *Yes*) the bureau had implemented each of the applications in the four areas, and (b) the perceived importance (0 = *Not important*, 1 = *Important*) of each of these applications in their organization's Web marketing efforts.

In order to measure the effectiveness of all the applications in each of the four functions, a 2 × 2 matrix was constructed (seeFigure 2). Using this matrix, a set of four possible scenarios was recognized. Quadrant I describes a bureau that does not use the application but perceives it important to its Web marketing efforts. If, on the other hand, the bureau is using the application and perceives the application to be important to its Web marketing strategy (Quadrant II), a practice of "effective" use of the application has been observed. Quadrant III was characterized as "wasting resources" because the bureau is using the application but does not perceive it to be important to its Web marketing strategy. Last, Quadrant IV was labeled "indifferent" whereby the bureau is not using the application and does not perceive it to be important to its overall Web marketing strategy. A scoring scheme was developed for all the applications in each of the four functions based on the four scenarios: a score of '+1' was assigned to Quadrant II (Effective), and a score of '−1' was assigned to both Quadrant I (Missing opportunities) and Quadrant III (Wasting resources) since both of them represent ineffective use of the applications. A score of '0' was assigned to Quadrant IV (Indifference) because it will not hurt the organization if it is not using the applications.

## 3.2 Dependent and Independent Variables Used in the Testing Model

The dependent variables in this study were: (a) the effectiveness of each of the four Web-based marketing functions (i.e., information, communication, transaction and assurance), and (b) the overall effectiveness of web-based marketing activities. An index was created to measure the effectiveness of each of the four functions by summing up all the scores in the four scenarios. For example, once an effectiveness index was calculated for each of the 12 applications of the information function, the indexes of the 12 applications were added to calculate the total effectiveness score of the information function. This same process was followed for all the applications and functions. The effectiveness scores for each function (i.e., information, communication, transaction, assurance) and the overall effectiveness of Web marketing activities are the dependent variables of the study. The overall effectiveness of Web-based marketing activities was calculated by adding up the effective indexes of the four marketing functions.

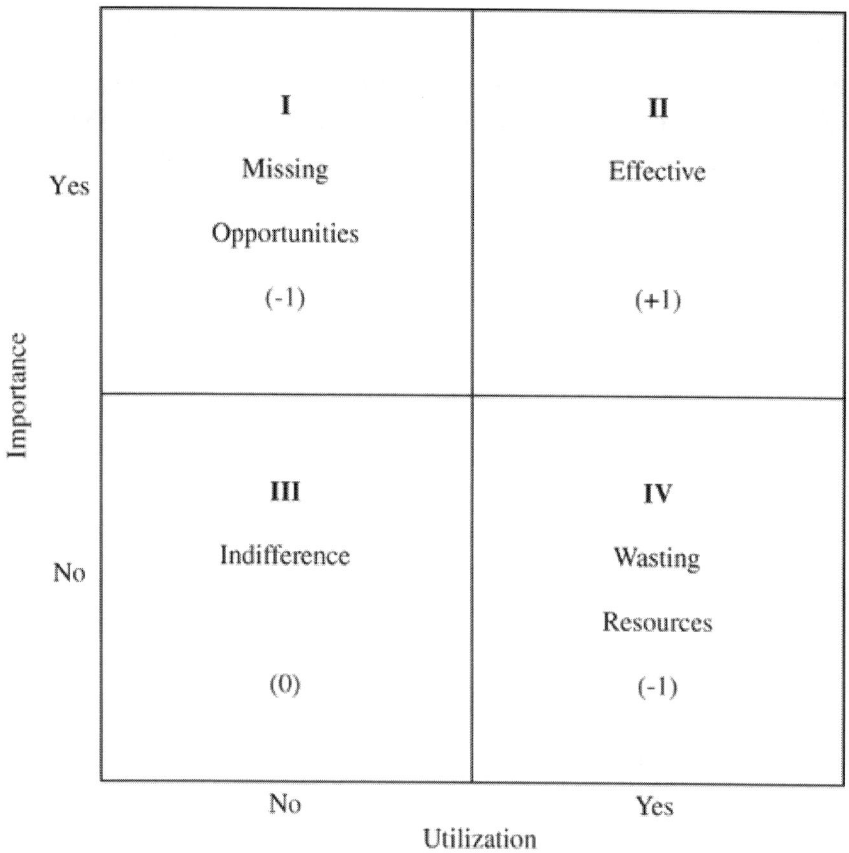


**Figure 2.** Effective evaluation matrix for technology applications in DMO web sites.

The independent variables being studied were the three groups of organizational variables: the technological environment of the organization, organizational characteristics, and the organizational technological competence. The first group of variables, the technological environment, included the availability of technology training programs, Web site budget, management involvement and support for technology operations. The second group of variables, the organizational characteristics, included two variables: organizational size measured by the CVB's yearly budget and number of full time employees, and the organizational innovativeness determined by the director's level of innovativeness. The measure of the latter variable was based on previous studies which show that an organization's innovativeness is mostly defined by the level of innovativeness of its management team, CEO or director (Thong, 1999). Several studies have found that an organization's management or leadership team has great impact on their strategic direction, innovation adoption and system implementation (Frambach, 1993; Main, 2002; Scupola, 2003; Thong, 1999). The third group of variables, the organizational technological competence, included management's and employees' technological knowledge measured by their rate of knowledge/skills with Internet technology. Table 1 shows the independent variables and their measures. A 7-point Likert scale was used with different anchors as noted in the table.

## 3.3 Sampling Frame and Data Collection

Convention and visitors bureaus at three levels (i.e., regional, county, and city) in the United States were used as the population of this study. The sample was drawn from a database constructed from the integration of various sources. Specifically, names of CVBs were obtained through several searches of the Internet using keyword searches including the names of each state (i.e., Indiana, New York, Wyoming, etc.), tourism, travel, and visitor centers. In addition, the

Web sites for each state were searched for up-to-date lists of CVBs. The results of these efforts were combined with a membership list provided by the International Association of CVBs. A total of 1,200 CVBs were identified and subsequently contacted using a brief telephone call to confirm their address and the name of the CEO/Director.

**Table 1.** Independent Variables and Measures

| Independent variables | Measure |
|---|---|
| Technology environment | Level of training |
| IT training | (1 = No training at all; 7 = Regular training) |
| Financial resources | Size of Web site budget |
| | (Dollar amount of Web site budget) |
| Management involvement | Level of management involvement |
| | (1 = Not at all involved; 7 = Extremely involved) |
| Management support | Level of management support |
| | (1 = Not at all supportive; 7 = Extremely supportive) |
| Organizational characteristics | Size of yearly budget |
| | (6 levels of ordinal measure) |
| Organizational size | Number of full time employees |
| | (6 levels of ordinal measure) |
| | Level of risk-taking in project |
| | (1 = Low risk; 7 = High risk) |
| | Level of reaction to outside changes |
| | (1 = Gradual and moderate; 7 = Aggressive and far reaching) |
| Organizational innovativeness | Timing of introduction of changes |
| | (1 = After competitors; 7 = Before competitors) |
| | Attitudes towards innovation |
| | (1 = Time tested methods; 7 = Innovative) |
| Technological competence | Level of management team's expertise |
| | (1 = Novice; 7 = Expert) |
| Management team's technological expertise | |
| Employee's technological expertise | Level of employee's technological expertise |
| | (1 = Novice; 7 = Expert) |

Using the Statistical Package of Social Sciences (SPSS) randomization procedure, the CVB list was then randomly divided into two groups with each consisting of approximately 600 CVBs, whereby one group was chosen as the sample frame for this study and the other group was used for a different study. This decision was made based on the resources available to the study as well as sound and calculative statistical reasoning. Based on previous working experience with CVBs and the representativeness of the sample, it was expected that the 600 CVB pool will generate a reasonable sample size which will allow sound and robust statistical analysis. The survey questionnaire was then mailed to the CEOs/Directors of 600 CVBs with a cover letter explaining the purpose of the survey and a request of assistance and support from the tourism organizations. Two follow up mailings at an interval of two weeks were sent to those who did not respond. A free copy of the executive summary of the study results was provided as an incentive to responding to the survey. A total of 268 CVBs returned the survey, among which 260 of the responses were found usable, representing a 43% response rate.

## 3.4 Data Analysis

The 260 usable surveys were analyzed by SPSS. First, descriptive statistics were performed on selected variables to obtain ranges, frequencies, and means. Second, regression analyses were conducted in order to examine the impact of the eight independent variables on each of the five dependent variables. The testing model for the regression analysis is presented in Figure 3.

Three independent variables were transformed to meet the requirement of regression analysis: financial resources, organizational size, and organizational innovativeness. Web site budget was used as a proxy measure for financial resources to support technology applications. A logarithm function of the original variable was applied to meet the normal distribution of the data associated with this variable. The mean of two variables, the total number of full time employees (six categories of ordinal measure) and CVB yearly budget (also six categories of ordinal measure), was used to measure organization size. Organizational innovativeness was measured by using the mean of four variables assessing the CEOs'/directors' project risk levels, reactions to outside changes, introduction of changes compared to competitors, and attitude to innovations. Five regression analyses were conducted to examine the impact of the technological environment, organizational characteristics and technological competence on each of the four stages as well as the overall effectiveness of Web marketing activities.

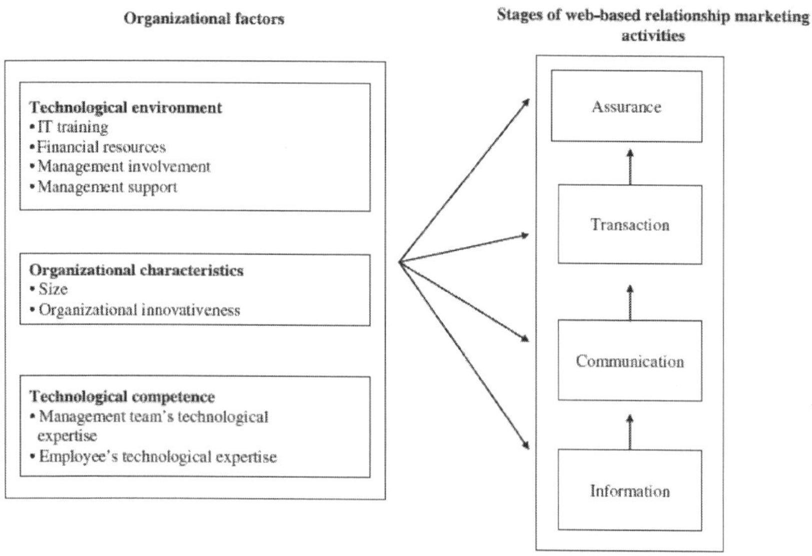


**Figure 3.** Testing model for impact of organizational factors on DMOs Web-based relationship marketing activities.

## 3.5 Study Results

Three levels of CVBs are represented in the sample group. The majority of them are county level (46%) and city level (45.6%) tourism offices, with only 8.4% representing regional level tourism offices. More than half of the tourism offices (51.6%) are represented by independent organizations. The others can be classified as: division of the Chamber of Commerce (22%), part of the county government (11.6%), and part of city government (9.6%).

The most important markets for the participating CVBs include the leisure market (90.3%), followed by the meetings/conventions market (63.7%) and the business travel market (40.9%). The distribution of the yearly budget presents a mixed picture. On the one hand, nearly 60% of the CVBs report that they have a yearly budget of less than $500,000, but on the other hand, quite a number of them (28.8%) have a yearly budget of more than one million dollars. This indicates that, measured by operating budgets, the majority of the CVBs are small and medium sized organizations. This assessment is confirmed by the number of full time employees: 90% of CVBs report less than 19 full time employees.

## 3.6 Assessment of Web-Based Relationship Marketing Applications

The next data analysis process involved the effectiveness of the applications under each of the four functions, calculated using the 2 × 2 effectiveness matrix discussed in the previous section. The effectiveness matrix provided four possible scenarios: missing opportunities (not used but important); effective (used and important); wasting resources (used but not important); and indifference (not used and not important). Table 2 presents the effectiveness percentages per quadrant for each application. The results are organized based on the effectiveness quadrant in decreasing order.

In the information function, 8 of the 12 applications were found to be effective (used and important): activities/attraction information (98.8%), accommodation information (98.5%), events calendar (96.2%), restaurant information (87.7%), shopping information (86.9%), links to regional/city/area pages (82.7%), maps/driving directions (79.6%), and travel guides/brochures (70%). However, providing virtual tours (67.7%) and tour operator information (41.5%) were found to be two applications where opportunities were being missed. This was because the functions were not being provided in the Web site but were considered to be important to the success of Web-based marketing. Furthermore, the analysis found CVBs were indifferent to providing banner advertisements (68.1%) and frequently asked questions (56.9%) in their Web sites.

In the communication function, only the brochure request capabilities (87.3%) application was found to be effective. Furthermore, the search functions application received mixed results splitting between the effective (48.1%) and missing opportunities (41.5%) quadrants. This demonstrates that CVBs have mixed feelings and perceptions about the use and importance of the application. However, trip vacation planner (61.2%) and interactive tools (56.2%) were found as missing opportunities. An application which received a high percentage on the indifference quadrant was the community functions (87.3%).

The results for the transaction function provide some alarming results since most applications fall under the indifference quadrant. Only one application, online reservations, was found to be a missing opportunity with 60.8% of the CVBs stating that it is not used and is important. On the other hand, CVBs stated that they were indifferent to the remaining six applications: Web seal certification (81.2%);

event tickets (73.8%); attraction tickets (73.1%); shopping carts and payment system (73.1%); secure transactions (70%); and themed products (68.1%). These functions were perceived by the CVBs as not used and not important to the Web marketing strategy. The result is in line with the purpose of CVBs, that is, they are destination marketing organizations but not sales focused organizations.

The results for the assurance function also reveal some areas of concern. Under the assurance function, five out of the eight functions were found to be missing opportunities: customer loyalty programs (70%); incentive programs (64.6%); personalization/customization (58.5%); direct e-mail campaign (54.2%); and e-mail newsletters (53.8%). This is an important observation since most CVBs recognize that they are missing opportunities by not providing these important applications in their Web sites. Furthermore, CVBs were indifferent (not used and not important) to the remaining three applications: cross-selling/up-selling opportunities (64.2%); privacy policy (59.2%); and highlight special offers/best buys (53.8%).

## 3.7 Results of Regression Analysis

Standard multiple regression analysis was selected for the study since it allows for a more sophisticated exploration of the interrelationships among a set of variables. Five regression models were formulated to examine the impact of the organizational factors on the Web-based marketing activities, with the following dependent variables for each of the five models: effectiveness of the information function (INF, Model I); effectiveness of the communication function (COM, Model II); effectiveness of the transaction function (TRA, Model III); effectiveness of the assurance function (ASS, Model IV); and overall effectiveness of Web-based marketing (OVERALL, Model IV). The same set of the independent variables were used in each of the five regression models: training programs (TP), financial resources (FR), management involvement (MI), management support (MS), organizational size (OS), organizational innovativeness (OI), management team's technological expertise (MKNOW), and employee's technological expertise (EKNOW). The multiple regression analyses yielded some interesting results (see Table 3).

Model I examines the influence of organizational factors on the effectiveness of information function. In formula:

$$INF = f\ (TP, FR, MI, MS, OS, OI, MKNOW, EKNOW)$$

Results for Model I revealed that 18.1% of the variance of this function is explained by the eight independent variables. The results of the standardized coefficients revealed that size of the organization ($\beta = .232$, sig. = .016) has the greatest impact on the effective use of information function, followed by management team's technological expertise ($\beta = .218$, sig. = .021). This means that larger CVBs are more successful in implementing the information function as part of their Web marketing strategy; but at the same time, the successful implementation of the information function has to be supported by management team's technological expertise.

Model II focuses on the influence of organizational factors on the effectiveness of communication function. In formula:

$$COM = f\ (TP, FR, MI, MS, OS, OI, MKNOW, EKNOW)$$

**Table 2**. Assessment of Effectiveness of Web Technology Applications Based on Functions

| Functions and applications | I Missing opportunities (%) | II Effective (%) | III Indifference (%) | IV Wasting resources (%) |
|---|---|---|---|---|
| Information | | | | |
| Activities/attraction information | 0.8 | 98.8 | 0 | 0.4 |
| Accommodation information | 0.8 | 98.5 | 0 | 0.8 |
| Events calendar | 2.3 | 96.2 | 0.4 | 1.2 |
| Restaurant information | 5.4 | 87.7 | 0.4 | 6.5 |
| Shopping information | 8.1 | 86.9 | 2.3 | 2.7 |
| Links to regional/city/area pages | 10.0 | 82.7 | 1.9 | 5.4 |
| Maps/Driving directions | 17.3 | 79.6 | 1.5 | 1.5 |
| Travel guides/brochures | 6.2 | 70.0 | 16.5 | 7.3 |
| Tour operator information | 41.5 | 48.8 | 7.7 | 1.9 |
| Frequently asked questions | 21.5 | 20.0 | 56.9 | 1.5 |
| Banner advertisements | 12.7 | 14.6 | 68.1 | 4.6 |
| Virtual tours | 67.7 | 11.5 | 18.5 | 2.3 |
| Communication | | | | |
| Brochure request capabilities | 9.2 | 87.3 | 1.5 | 1.9 |
| Search functions | 41.5 | 48.1 | 6.9 | 3.5 |
| Interactive tools | 56.2 | 28.1 | 15.4 | 0.4 |
| Trip/Vacation planner | 61.2 | 23.1 | 15.4 | 0.4 |
| Community functions | 11.9 | 0.4 | 87.3 | 0.4 |
| Transaction | | | | |
| Online reservations | 60.8 | 20.0 | 18.5 | 0.8 |
| Themed products | 13.1 | 17.3 | 68.1 | 1.5 |
| Secure transactions | 19.6 | 9.6 | 70.0 | 0.8 |
| Event tickets | 16.5 | 9.2 | 73.8 | 0.4 |
| Attraction tickets | 18.1 | 8.5 | 73.1 | 0.4 |
| Shopping carts & payment system | 18.5 | 7.3 | 73.1 | 1.2 |
| Web seal certification | 16.5 | 1.9 | 81.2 | 0.4 |
| Assurance | | | | |
| E-mail newsletters | 53.8 | 33.5 | 11.5 | 1.2 |
| Highlight special offers/best buys | 13.1 | 30.4 | 53.8 | 2.7 |
| Direct e-mail campaign | 54.2 | 28.8 | 15.8 | 1.2 |
| Personalization/Customization | 58.5 | 22.7 | 17.7 | 1.2 |
| Privacy policy | 18.5 | 18.8 | 59.2 | 3.5 |
| Incentive programs | 64.6 | 17.7 | 16.9 | 0.8 |
| Cross-selling/Up-selling opportunities | 20.0 | 13.8 | 64.2 | 1.9 |
| Customer loyalty programs | 70.8 | 3.8 | 25.0 | 0.4 |

**Table 3.** Results of Regression Analysis

| Independent variables | Model I information | | Model II communication | | Model III transaction | | Model IV assurance | | Model V overall | |
|---|---|---|---|---|---|---|---|---|---|---|
| | Betas | Sig. | Betas | Sig. | Betas | Sig. | Betas | Sig. | Betas | Sig. |
| IT Training | .082 | .248 | .101 | .125 | .019 | .805 | .051 | .464 | .087 | .172 |
| Financial resources | .061 | .524 | .254 | .005*** | .138 | .176 | .249 | .009*** | .241 | .006*** |
| Management involvement | .096 | .305 | .093 | .287 | .056 | .575 | −.036 | .694 | .065 | .444 |
| Management support | −.068 | .475 | −.135 | .123 | .012 | .905 | .017 | .858 | −.053 | .532 |
| Size | .232 | .016** | .210 | .018** | .131 | .196 | .156 | .096* | .254 | .003*** |
| Organizational innovativeness | .090 | .230 | .076 | .271 | −.022 | .785 | .066 | .365 | .078 | .246 |
| Management team's technological expertise | .218 | .021** | .182 | .037** | −.028 | .781 | .138 | .132 | .186 | .028** |
| Employee's technological expertise | −.078 | .355 | −.001 | .994 | .057 | .523 | .027 | .747 | −.002 | .975 |
| $R^2$ | 18.1% | | 29.8% | | 7.7% | | 21.6% | | 33.9% | |

*Note.* *$p < .10$; **$p < .05$; ***$p < .01$.

Results for Model II revealed that 29.8% of the variance of the function is explained by the independent variables. The results of the standardized coefficients revealed that financial resources (β = .254, sig. = .005), organizational size (β = .210, sig. = .018), and management team's technological expertise (β = .182, sig. = .037) were found to be statistically significant in explaining the effectiveness of the communication function. This means that CVBs that seek to implement the communication function in their Web marketing strategies will need to be aware of the importance of sufficient allocation of financial resources in supporting technology strategies. In addition, similar to the information function, organizational size and management team's technological expertise can also play important roles in the effective implementation of communication functions.

Model III examines the influence of organizational factors on the effectiveness of the transaction function. In formula:

$$TRA = f\ (TP, FR, MI, MS, OS, OI, MKNOW, EKNOW)$$

Results of Model III revealed that only 7.7% of the variance of the function is explained by all the independent variables, which means that these independent variables have little power in explaining the variance of the transaction function. Consistent with the poor overall model fit, the standard coefficients revealed that none of the independent variables were found to be statistically significant in explaining the variance of the dependent variable. This can probably be explained by two observations. First, most of the CVBs in North America context are legally constrained in providing transaction capabilities in their Web sites so the effectiveness of this function is not even an issue in their marketing agenda. Second, CVBs in North America are often chartered as the destination marketing organization. As a result, they position themselves as marketing organizations but not sales agencies.

Model IV examines the influence of organizational factors on the effectiveness of the assurance function. In formula:

$$ASS = f\ (TP, FR, MI, MS, OS, OI, MKNOW, EKNOW)$$

Results from Model IV revealed that 21.6% of the variance of this function is explained by the independent variables. The results of the standardized coefficients revealed that financial resources ($\beta = .249$, sig. = .009) and organization size $\beta = .156$, sig. = .096) were found to be statistically significant in explaining the effectiveness of the assurance function. CVBs seek to implement the assurance function within their Web marketing strategies, which need to be supported by sufficient allocation of financial resources, a variable which is usually positively correlated to organizational size.

Model V focuses on the influence of organizational factors on the overall effectiveness of Web-based marketing activities. In formula:

$$OVERALL = f\ (TP, FR, MI, MS, OS, OI, MKNOW, EKNOW)$$

The results of Model V revealed that 33.9% of the variance of this function is explained by the independent variables. The results of the standardized coefficients revealed that organizational size ($\beta = 0.254$, sig. = .003), financial resources ($\beta = .241$, sig. = .006) and management team's technological expertise ($\beta = .186$, sig. = .028) were found to be statistically significant in explaining the overall effectiveness of Web-based marketing activities.

The results of the multiple regressions revealed that across all the independent variables affecting the five dependent variables, the most significant organizational factors are organizational size, management team's technological expertise, and financial resources. It can be concluded from the analyses that bigger organizations with more financial resources and management with technological expertise are important to the adoption and successful implementation of the different functions in their Web-marketing efforts. The results of the study suggest that bigger organizations have more financial resources available to implement technology innovations and that technology savvy leadership is helpful and sometimes imperative to support and implement Web-based marketing strategies.

## 4. DISCUSSION AND CONCLUSIONS

This study aimed to assess the Web-based destination marketing activities employed by American CVBs. Previous research has demonstrated that many organizational factors impact the way organizations adopt and implement IT innovations (Thong, 1999; Tornatzky & Fleischer, 1990; Zhu & Kraemer, 2005). This study provides an additional piece in the puzzle to better understand the use of Web site applications for the successful execution of Web marketing strategies from a relationship marketing perspective. The study examined the use and effectiveness of Web applications by American CVBs and the impact of organizational factors on the level of Web functions implemented. Multiple regression analyses found that size, financial resources, and management team's technological expertise have the most impact on the effectiveness of Web marketing functions implemented by CVBs.

First, the results of the study revealed that most CVBs use their Web sites mainly for information provision purposes with less focus placed on the communication, transaction and assurance applications. This finding by and large supports the findings of previous research that Web sites are used mainly to provide information to the consumer and not for transaction or relationship building purposes (Dore & Crouch, 2003; Palmer & McCole, 2000). The significance of this finding is highlighted by the fact that CVBs need to understand the importance of the Internet and the purpose of their own Web sites and establish its potential value compared to more traditional promotional activities (So & Morrison, 2003). In other words, CVBs need to think about and learn how they can use their Web sites to fully practice relationship marketing. It is possible that their managers need to be educated both in the Web-based marketing and relationship marketing. They should be educated that the essence of Web-based marketing is not about technology or technology applications; rather, the focus should be on how to use technology or technology application to organize and manage information in order to achieve the ultimate goal of managing customer relationships.

Second, the study examined the effectiveness of the Web applications. The CVBs participated in this study reported that applications categorized under the information function are perceived to be the most effective. These applications are categorized as effective since they are used and perceived to be important for the successful implementation of Web marketing strategies. On the other hand, applications under the communication, transaction and assurance functions were found to be either lost opportunities or indifferent to the overall success of the Web-based marketing agenda. These results show that there are still great opportunities within the Web marketing domain CVBs can take advantage of in order to implement Web applications to create and retain relationships with consumers.

Third, the findings of the multiple regression analyses both support and challenge the results of previous studies. This study focuses on CVBs, which are unique organizations in that they are chartered to promote a destination and its services without any monetary gain for the organization. The organization's size, financial resources and management team's technological expertise were found to have an impact on the functions implemented by a CVB. Model I (information function) showed that size and management team's technological expertise are the only two functions having an impact on this function. Model II (communication) found that financial resources, size and management team's technological expertise all have impacts, in different degrees, on the communication function. Model III (transaction) found that none of the variables identified have an impact on the transaction function. Model IV (assurance) found that both allocation of financial resources and size have impacts on the assurance function. Finally, Model V (overall effectiveness), which represents the combined functions, found that financial resources, organizational size, and management team's technological expertise all have great impacts on the overall effectiveness of Web marketing activities.

In relation to the presence of IT training programs, the study implies that it does not have an impact on the effectiveness of Web marketing functions implemented by a CVB. It can be argued that most people are familiar with the basic functions of the Internet and its use, therefore making the presence of IT training programs not necessary for their Web marketing strategies. This study further suggests that the allocation of financial resources was the variable that has the most significant impact on the Web marketing functions implemented by CVBs. Previous studies found that the allocation of financial resources for IT adoption and implementation has a positive impact on innovation adoption behavior (Yuan, Gretzel, & Fesenmaier, 2003). The results of this study support the findings of previous research stating that organizations that have more resources will be more likely to implement IT innovations and have a higher level of success in their implementations (Zhu & Kraemer, 2005). The research findings particularly imply that the financial resource variable has the most significant contribution to the implementation of the communication, assurance and the overall effectiveness of Web marketing functions. This issue should be specifically considered by senior executives of CVBs. However, small CVBs have a disadvantage here since they do not have substantial financial resources compared to larger CVBs. Their managers and executives need to use their resources in a more effective way. Again cooperating and working together with neighboring CVBs can be a option for them to combine their resources and knowledge to achieve cost efficiency in implementing Web-based marketing activities.

Size has been the most investigated organizational characteristics in relation to technology adoption and implementation (Tornatzky & Klein, 1982). Previous studies have found that the size of the organization has both a positive and negative impact to the adoption, implementation and extent of use of IT applications. However, this study found that size has a significant impact on the implementation of information, communication, assurance and the overall effectiveness of Web marketing functions. Tornatzky & Fleischer

(1990), Lefebvre & Lefebvre (1992), Frambach (1993), Thong (1999) found that size has a positive impact on the adoption, diffusion and extent of use of innovations, more specifically IT innovations. In line with this research, Wang & Fesenmaier (2006) found that size is positively related to the implementation of Web marketing strategies. However, Main (2002) and Zhu & Kraemer (2005) found that size has a negative impact on the extent of e-business use. Furthermore, Goode and Stevens (2000) found that size is not related to http://www technology adoption. The results of this study may be contradictory to some previous findings because this research is based on self reported answers from CVB directors/CEOs. In addition, this type of organization is different from the typical for-profit business organizations. A CVB is typically a small organization with limited resources whose goal is to market a destination and its services to help develop the economy of the region. Given the positive relationship between size and resources, it is not surprising to find that size makes the most significant contribution to the implementation of Web marketing functions. This research finding implies that because of their size, small CVBs may face challenges in finding resources and subsequently implementing the Web-based destination marketing activities from the perspective of relationship marketing. As stated above, small CVBs can either work with other small CVBs or collaborate with larger nearby CVBs.

The influence of organizational innovativeness as measured by the CEOs' innovativeness was not found to have an impact on any of the Web functions. The lack of significant relationship between the variables may be due to several aspects such as top management's risk aversion, reaction to outside changes, or innovation adoption behavior. The various variables that impact CEO innovativeness are critical to the overall strategy, therefore making innovativeness a complicated variable to examine.

The study findings imply that management team's technological expertise has a strong impact on the implementation of several functions: information, communication, and overall effectiveness. The results of this study support the findings of previous research that management's knowledge and expertise are extremely important to the likelihood to adopt and implement IT innovations (Thong, 1999; Wang & Fesenmaier, 2006; Yuan et al., 2003). That is, a more knowledgeable management team will support the decision to adopt innovations and realize the benefits of those innovations for the organization (Thong, 1999, Yuan et al., 2003). The study also suggests that employee's technological expertise does not have an impact on the level of Web-based marketing functions. This may be due to the little or no influence employees have on the decision to adopt applications that support the different levels of Web marketing applications. In other words, the main decision to adopt an application usually rests with upper management, marginalizing the impact of employees on the decision related to Web marketing strategy.

The research results of this study will be of interest to CVBs, DMOs and other tourism offices. The findings can help shed light on the use, importance and effectiveness of Web applications. A better understanding of the use of Web applications will allow destination marketers to appropriately allocate resources to support those applications which are found to be important and effective for the overall Web marketing strategy. The study results clearly illustrate the role and impact of organizational factors on the different Web functions. A better understanding of the factors that have the greatest impact will help DMOs make appropriate decisions to improve the effectiveness of their Web marketing practices. DMOs should understand that technology is only a tool not the end game; its effective implementation depends on the right design of the organizational structure and capabilities. Given the fact that CVBs operate in information intensive, complex and competitive business environment, they are required to demonstrate their capabilities and competencies in designing and applying Web-based marketing activities following a relationship marketing logic. As discussed above, this is a challenging process which requires investment of considerable resources and organizational support for a long period of time in order to be able to see positive outcomes.

## 5. STUDY LIMITATIONS AND SUGGESTIONS FOR FUTURE RESEARCH

This study has a number of limitations. First, data was mainly collected at the county and city level CVBs in the US. In terms of operating budgets and full time employees, the majority of the participating CVBs were small and medium sized organizations. About 90% of CVBs participated in the study had less than 19 full time employees. Therefore, the study findings may not be relevant for large DMOs in the US and other countries. The most important markets for the participating CVBs were the leisure market (90.3%) and the meetings/conventions market (63.7%). It is possible that CVBs targeting other business segments may be operating differently. The study looked at how organizational characteristics influence implementation of Web-based relationship marketing activities. Certainly, there may be other factors influencing deployment of Web-based marketing activities. In addition, the data for this study was collected in the early 2000s. Given the rapid changes in IT and online marketing activities, the CVBs participated in the current study might have changed their IT applications and online marketing strategies.

The findings of this study both support and challenge the findings of previous studies. This demonstrates that more research is needed in this field to provide a better understanding of the impact of the organizational factors on technology applications for this type of organization. In addition, the topic of RM in the tourism field is greatly under researched and future studies in this area will create a better understanding of the topic and provide practical guidance to implementing such practices in daily operations in order to make their Web marketing practices more effective and efficient. In addition, further research is needed to investigate the influence of the organizational technology capability on the Web-based marketing functions and strategies. This will provide more clarity on the organizational factors needed to successfully implement a Web based marketing strategy focused on relationship building and retention. Future research should also include the consumer's perspective in understanding their use of destinations' Web sites. This will provide a comprehensive view and understanding of RM strategies through Web applications. For example, research examining the consumer's perspective on the aspect of Web-based RM can be conducted to identify the differences between consumers and organizations in terms of what is relevant. This comparison will provide valuable information on what applications are important and how CVBs and other tourism organizations can adjust its Web-based marketing strategies to fit the needs and preferences of the consumers.

## REFERENCES


1. Ahn, J. Y., Kim, S. K. and Han, K. S. 2003. On the design concepts for CRM system. *Industrial Management & Data Systems*, 103: 324–331.

2. Berry, L. L. 1995. Relationship marketing of services: Growing interest, emerging perspectives. *Journal of the Academy of Marketing Science*, 23: 236–245.

3. Buhalis, D. 1998. Strategic use of information technologies in the tourism industry. *Tourism Management*, 19: 409–421.

4. Buhalis, D. 2000. Marketing the competitive destination of the future. *Tourism Management*, 21: 97–116.

5. Contractor, N. S., Wasserman, S. and Faust, K. 2006. Testing multi-theoretical, multi-level hypotheses about organizational networks: An analytical framework and empirical example. *The Academy of Management Review*, 31: 681–703.

6. Damanpour, F. 1991. Organizational innovations: A meta-analysis of effects of determinants and moderators. *Academy of Management Journal*, 34: 555–590.

7. Davis, F. D., Bagozzi, R. P. and Warshaw, P. R. 1989. User acceptance of computer technology: A comparison of two theoretical models. *Management Science*, 35: 982–1003.

8. Ditto, S. and Pille, B. 1998. Marketing on the Internet. *Healthcare Executive*, 13: 54–55.

9. Doolin, B., Burgess, L. and Cooper, J. 2002. Evaluating the use of the Web for tourism marketing: A case study from New Zealand. *Tourism Management*, 23: 557–561.

10. Dore, L. and Crouch, G. I. 2003. Promoting destinations: An exploratory study of publicity programs used by national tourism organizations. *Journal of Vacation Marketing*, 9: 137–151.

11. Frambach, R. T. 1993. An integrated model of organizational adoption and diffusion innovations. *European Journal of Marketing*, 27: 22–41.

12. Goode, S. and Stevens, K. 2000. An analysis of the business characteristics of adopters and non-adopters of World Wide Web

technology. *Information and Technology Management*, 1: 129–154.

13. Gretzel, U., Yuan, Y. and Fesenmaier, D. R. 2000. Preparing for the new economy: Advertising strategies and change in destination marketing organizations. *Journal of Travel Research*, 39: 146–156.

14. Grönroos, C. 1990. Relationship approach to marketing in service contexts: The marketing and organizational behavior interface. *Journal of Business Research*, 20: 3–11.

15. Gummesson, E. 1994. Making relationship marketing operational. *International Journal of Service Industry Management*, 5: 5–20.

16. Hanson, W. 2000. *Principles of Internet marketing*, Mason, OH: South-Western College Publishing.

17. Kim, J., Suh, E. and Hwang, H. 2003. A model for evaluating the effectiveness of CRM using the balanced scorecard. *Journal of Interactive Marketing*, 17: 5–19.

18. Kotler, P., Bowen, J. and Makens, J. 2003. "Building customer loyalty through quality". In *Marketing For Hospitality and Tourism*, 3rd, Edited by: Kotler, P. 380–440. Upper Saddle River, NJ: Prentice Hall.

19. Lefebvre, E. and Lefebvre, L. A. 1992. Firms innovativeness and CEO characteristics is small manufacturing firms. *Journal of Engineering and Technology Management*, 9: 243–277.

20. Lefebvre, L., Mason, R. and Lefebvre, E. 1997. The influence prism in SMEs: The power of CEO perceptions on technology policy and its organizational impacts. *Management Science*, 43: 856–878.

21. Lexhagen, M. 2005. The importance of value-added services to support the customer search and purchase process on travel Websites. *Information Technology and Tourism*, 7: 119–135.

22. Main, H. C. 2002. The expansion of technology in small and medium hospitality enterprises with a focus on net technology.*Information Technology and Tourism*, 4: 167–174.

23. Palmer, A. and McCole, P. 2000. The role of electronic commerce in creating virtual tourism destination marketing organizations. *International Journal of Contemporary Hospitality Management*, 12: 198–204.

24. Ritchie, R. J. B. and Ritchie, J. R. B. 2002. A framework for an industry supported destination marketing information system.*Tourism Management*, 23: 439–454.

25. Rogers, E. M. 1995. *Diffusion of Innovations*, 4th, New York: The Free Press.

26. Scupola, A. 2003. The adoption of Internet commerce by SMEs in the South of Italy: An environmental, technological and organizational perspective. *Journal of Global Information Technology Management*, 6: 52–71.

27. Sharma, A. 2002. Trends in Internet-based business-to-business marketing. *Industrial Marketing Management*, 31: 77–84.

28. So, S. and Morrison, A. M. 2003. Destination marketing organizations' Web site users and nonusers: A comparison of actual visits and revisit intentions. *Information Technology and Tourism*, 6: 129–139.

29. Subramanian, C., Shaw, M. J. and Gardner, D. M. 2000. Product marketing and channel management in electronic commerce. *Information Systems Frontier*, 1: 363–378.

30. Thong, J. Y. L. 1999. An integrated model of information systems adoption in small businesses. *Journal of Management Information Systems*, 15: 187–214.

31. Tornatzky, L. G. and Fleischer, M. 1990. *The process of technological innovation*, Lexington, MA: Lexington Books.

32. Tornatzky, L. G. and Klein, K. J. 1982. Innovation characteristics and innovation adoption-implementation: A meta-analysis of findings. *IEEE Transactions on Engineering Management*, 29: 28–45.

33. Vesanen, J. and Raulas, M. 2006. Building bridges for personalization: A process model for marketing. *Journal of Interactive Marketing*, 20: 5–20.

34. Wang, Y. and Fesenmaier, D. R. 2006. Identifying the success of Web-based marketing strategy: An investigation of convention and visitors bureaus in the United States. *Journal of Travel Research*, 44: 239–255.

35. Wang, Y. and Russo, S. 2007. Conceptualizing and evaluating the functions of destination marketing systems. *Journal of Vacation Marketing*, 13: 187–203.

36. Yuan, Y., Gretzel, U. and Fesenmaier, D. R. 2003. Managing innovation: The use of Internet technology by American convention and visitors bureaus. *Journal of Travel Research*, 41: 240–256.

37. Zhu, K. and Kraemer, K. L. 2005. Post-adoption variations in usage and value of e-business by organizations: Cross-country evidence from the retail industry. *Information Systems Research*, 16: 61–84.

38. Zineldin, M. 2000. Beyond relationship marketing: Technologicalship marketing. *Marketing Intelligence & Planning*, 18: 9–23.

# CHAPTER 4

# Corporate Marketing Objectives and Evaluation Measures for Integrated Television Advertising and Sports Event Sponsorships

**Peter Smolianov & Joseph F. Aiyeku**

Salem State College, Salem, Massachusetts, USA

## ABSTRACT

To determine how integrated TV advertising and event sponsorship should be evaluated, a theoretical framework derived from global exploratory research of academic literature and consulting reports of more than 50 authors was validated by 16 experts. To investigate how evaluation was managed in practice, 12 campaigns which had sponsored a televised event and placed advertisements during the broadcast of this event were analyzed via case studies. The investigated competitions included the Wimbledon Tennis

Tournament in London and the Olympic Games in Sydney. The examined brands comprised automotive, financial services, retail chain, office equipment, and consumer goods. The study identified the key techniques that lead to increased corporate sales—four objectives and necessary performance measures, which could be useful to all executives promoting brands through televised sport.

## KEYWORDS

advertising, corporate sponsorships, event marketing, integrated marketing communications, Olympics, sport, television, Wimbledon Tennis Tournament

## 1. INTRODUCTION

Mansourpour (2007) points out "Sports sponsorship is a tool used increasingly by corporations to generate awareness, alter attitudes and attempt to influence consumer behaviour patterns. It is able to cut through clutter effectively, target specific consumer segments, and to generate beneficial consumer effects. These aspects have meant that as an element of the promotional mix, sports sponsorship is becoming an important practice to a wider selection of firms."

But at a time of increasingly competitive and sophisticated corporate marketing through televised sporting events and a growing concern about the return on investment it provides (Lawson, 2002), the management and evaluation of corporate event sponsorship and advertising through sport still requires better understanding (Olkkonen, 2001; Farrelly, 2002;Cliffe & Motion, 2005). A number of authors have suggested that one best practice was to "integrate" event sponsorship with advertising, and indicated that further research was needed to investigate how such integration should be managed and evaluated (Quester & Thompson, 2001; Farrelly, 2002; Shimp, 2003). More recently, Sylvestre and Moutinho (2008) indicated sports sponsors are viewed as being more effective in conveying messages, not only because of the size and variety of the audience, but because of the resources that they have at their disposal.

Some companies currently pay tens of millions of dollars for the sponsorship of one major televised event, e.g. the Olympic Games or the Australian Tennis Open (Mellish, 2000; Stensholt, 2003; Spais & Filis, 2006a, 2006b), and they also spend tens of millions on television advertising during these events, e.g., the Olympic Games or the annual Super Bowl football championship game in the United States (Mellish, 2000; Collins, 2003; O'Reilly, McCarthy, Séguin, & Lyberger, 2005;Papadimitriou, Apostolopoulou, & Dounis, 2008).

Despite the increasing importance of integrated sponsorship and advertising as a tool for corporate marketing, conventional TV advertising methods and their evaluation are more developed than the corresponding event sponsorship practices (Shimp, 2000; Farrelly, 2002; Lane, King, & Russell, 2008; Papadimitriou et al., 2008). An estimated 70% of traditional broadcast media campaigns are evaluated, whereas only 28% of sponsorships are measured (*S:COMM Research*, 2003). The problem of effective evaluation of such campaigns is a key part of the present study.

Many best practice marketing campaigns now link advertising and sponsorship messages (Shannon, 1999; Stamler, 2002). For example, advertising and sponsorship by Fosters using the AFL 2000 Grand Final of Australian Rules football were integrated around the company's positioning message—"I believe." During commercial breaks in the game broadcast, Fosters—a beer producing company—showed commercials which utilized football images bonded with the event and, together with football-related messages, used the corporate positioning tagline: "I play football without a helmet. I believe." The message was followed by the corporate logo as it appears on Fosters beer bottles. The same messages and the logo were displayed on the most exposed spots of the event ground—behind the gates, and behind the goal umpire on whom cameras focused as he registered each goal.

Meenaghan (2002) pointed out that the communication objectives of advertising and sponsorship are similar. The two media are in some respects interdependent, and when integrated, the effectiveness of both relies on the support of each other. Just like advertising campaigns, sponsorships of televised events have to be carefully designed, planned, executed, and measured (Rossiter & Percy, 1997; Farrelly, 2002). This is particularly important when sponsorships and advertising are integrated, as both can share design, execution, and measurement.

The literature indicated that the different strengths of advertising and sponsorship should be used to achieve common objectives of the two media (Meenaghan, 2002). The persuasive strength of sponsorship lies in its ability to convey the commercial message and influence the consumer in a more voluntary fashion than the standard advertising sell. The corporate message can be received in a more positive light when the company responsible is seen as a supporter of the sporting event that the individual values. This effect may be particularly strong among followers of the sponsored event and sport. However, successful translation of this "positive light" or "rub-off" from images of sporting events, stars, and properties into an enhanced sponsor's brand image and the achievement of other corporate objectives necessitates effective integration of sponsorship with conventional advertising (Farrelly, 2002).

The strength of a conventional advertisement is in its propensity to send a direct and specific message. Sponsorship of an event, on the other hand, gives an opportunity to indirectly deliver a message, e.g., increase brand awareness and enhance brand image which could lead to raised sales. Also, in non-sales environments (e.g., the "Quit smoking" campaign) both media can contribute to changing the attitudes and behavior of sports viewers. The unique properties of the two media offer synergistic opportunities that are not available when employed separately.

It was clear from the literature that the major event sponsorship investment is reliant on support from other forms of promotions, primarily television advertising. It was also clear that sponsorship should be developed and evaluated together with advertising. However, it yet has to be explained how the integration should be achieved, in terms of management and evaluation. This study attempted to discover the criteria that can be used for objectives and measures of integrated advertising and sponsorship and how practitioners actually use these criteria.

## 2. THEORETICAL FRAMEWORK

A number of authors agreed that general advertising objectives and measures should be employed for major event sponsorship (Farrelly, 1995, Hansen & Scotwin, 1995; Arthur, Scott, Woods & Booker, 1998; Rossiter, 1999; Farelly, 2002;Meenaghan, 2002). In essence, these authors recommended that sport sponsorship objectives and measures (like advertising objectives and measures) be based on consumer behavior steps. The following steps guide the consumers along the way from being exposed to the promotional message to changing attitudes and behavior (Rossiter & Percy 1997;Percy, Rossiter & Elliott, 2001; Belch & Belch, 2009):

1 Exposing the target market to the promotional message through media coverage is an important advertising and event sponsorship objective and measure (Meenaghan, 2002; Belch &

Belch, 2009). Examples of exposure-related objectives are increased target market via television coverage (Irwin & Sutton, 1994; Arthur, Woods & Scott, 2000); audience profile and size determined and targeted (Arthur et al., 2000); media exposure or increased media exposure (Farrelly, 1995;Cheng, 2000); audience demographic fit (Grdovic, 1992); cost per target market (Grdovic, 1992); gain in media coverage (Grdovic, 1992); exposure in seconds on television (Hansen & Scotwin, 1995); TV as a cost-efficient alternative exposure (Hirons, 1990); an opportunity to reposition a brand to alter target market (Hirons, 1990); demographic profile of immediate or extended audience (Irwin, Assimakopoulas & Sutton, 1994); overcoming communication barriers by targeting markets in non-commercial situations (Kuzin & Kutepov, 1994); increased frequency and quality of contacts with target markets (Kuzin & Kutepov, 1994); TV or media coverage (McDonald, 1991); level of media coverage and exposures gained (Meenaghan, 1983, 1991, 2002); coverage of the target audience (Meenaghan, 1983, 1991, 2002); increased television, press, and radio coverage/exposure (Quester, 1995; Shilbury, Quick & Westerbeek, 2003); and targeting new market segments (Shilbury et al., 2003).

2 Making target markets aware of the promotional message, sponsoring company, brand, and product are important advertising and event sponsorship objectives (Hoek, 1999; Rossiter, 1999; Hal Dean, 2002; Meenaghan, 2002;Berkes & Nyerges, 2004; Berkes, Nyerges, & Váczi, 2008; Belch & Belch, 2009). Awareness-related goals comprise a linking of the sponsor's name to the property (Arthur et al., 2000); creation of brand awareness or an increase in brand or product awareness (Hirons, 1990; Kuzma, 1990; Irwin & Sutton, 1994; Kuzin & Kutepov, 1994; Farrelly, 1995; Quester, 1995; Cheng, 2000; Shilbury et al., 2003); awareness of the event, awareness that the event is sponsored, awareness of the event sponsor (Grdovic,

1992); pre- and post-event product awareness (Grdovic, 1992), increased company awareness (Kuzma,); awareness as a possible sponsorship measurement point (Meenaghan, 2002); and increased public and target market awareness (Thwaites & Carruthers, 1998). The study's theoretical framework takes into account the fact that event sponsorship makes a wider audience aware of the promoted brand, targeting live event audiences as well as TV viewers (S:COMM Research, 2003). Effective integration may allow TV advertising to increase brand awareness among live audiences who also watch TV broadcasts, reinforcing the messages displayed at the venue.

3 Changing or enhancing the brand image in target markets' minds and developing a disposition to change behavior towards buying the promoted brand is the basic reason for event sponsorship (Rossiter, 1999). The image-related objectives and measures include "clean image" (McDonald, 1991; Arthur et al., 2000); "improved" or "enhanced" company or product image (Hirons, 1990; Kuzma, 1990; Grdovic, 1992; Irwin & Sutton, 1994; Smolianov, 1994; Farrelly, 1995;Quester, 1995; Thwaites & Carruthers, 1998; Hal Dean, 2002; Meenaghan, 2002; Smolianov & Shilbury, 2005); an image to provide a company/market fit (Irwin et al., 1994); a "changed" image (Kuzin & Kutepov, 1994); an image as a possible sponsorship measurement point (Meenaghan, 2002); an image creation and improvement (Shilbury et al., 2003); and enhancement of the company's image with employees (White & Mazur, 1995). Both event sponsorship and advertising contribute to an enhanced brand image (Rossiter, 1999), but sponsorship functions in a different manner to advertising and other forms of promotion (Meenaghan, 2001).

4 As Meenaghan (2001) argued, sponsorship engages the consumer by the fact that the event sponsored in sports can be one with which the consumer has an intense emotional relationship, a quality which is apparent in sports fans and sport

enthusiasts generally. This intensity of relationship of the consumer with the sport can help explain the concepts of goodwill, perceptions of benefit, and image transfer, which can result in a positive consumer response to the brand or product. Image transfer alone represents an important sponsorship objective (Meenaghan, 2001). Apart from influencing target sports audiences who are TV viewers, the sponsorship of the event can also affect the broader audience of the sponsored activity (Meenaghan, 2002). Sponsership may include the sponsor's employees and business clients targeted through event hospitality guests programs and staff participation programs as part of event sponsorship (Meenaghan, 1983, 1991; Farrelly, 1995; White & Mazur, 1995; Thwaites & Carruthers, 1998; Arthur et al., 2000).

5 Changing target markets' behavior, such as increasing use or sales of the promoted brand is the ultimate aim of advertising and sponsorship (Hoek, 1999; Meenaghan, 2002). The sales-related objectives and measures include increased sales, and short-term and long-term sales (Farrelly, 1995; Grdovic, 1992; Kuzma, 1990; Smolianov, 1994;Quester, 1995; Arthur et al., 2000; Shilbury et al., 2003; Smolianov & Shilbury, 2005); sales in general (Meenaghan, 1983,1991; Thwaites & Carruthers, 1998; Cheng, 2000; Meenaghan, 2002); improvement in market share (Irwin & Sutton, 1994;Hansen & Scotwin, 1995); generation of sales (Grdovic, 1992); support product sales, increased sales among mass consumer groups by promoting products through athletes who influence public opinion (Kuzin & Kutepov, 1994); increased sales among recreational users (Kuzin & Kutepov, 1994); quantifiably increased sales (Stotlar, 1998); and benefits of social marketing (O'Reilly & Madill, 2007). Without the incorporation of behavioral measures such as sales or brand usage into the study's sponsorship and advertising management framework, evaluation would remain limited to an

investigation of awareness, attitudes (image), and possibly preferences, with no understanding of the relationship between these variables and consumer behavior. According to Quester and Thompson (2001), adoption of an evaluation framework based on behavioral measures could help to minimize the possibility that investment in sponsorship is simply a magnanimous but empty gesture.

# 3. METHOD

The methods of this study were determined taking into account the suggestions made by Farrelly (1995), Meenaghan (1999) and Olkkonen (2001). Meenaghan (1999) recommended investigating corporate sport marketing practices by using a qualitative methodology, noting that many previous studies were based on surveys of sponsorship decision makers. Meenaghan was of the opinion that these replications meant declining returns in terms of improved understanding, and that usage of highly structured survey instruments imposed a researcher's view of practice. Farrelly (1995) indicated the need for both qualitative and exploratory research, arguing that the lack of flexibility in previous survey methodology resulted in some conflicting evidence about sport marketing and sponsorship management. Olkkonen (2001) concluded from his analysis of literature that most of current sponsorship research shows a need for further development of theoretical frameworks and research designs that would allow a deeper, action-oriented understanding of this field of interest.

The measures identified through exploratory research and theoretical reasoning were validated through discussions with experts to modify the framework and to finalize the questions for interviews with practitioners.

## 3.1 Expert Validation

Validation of the theoretical framework was conducted through two series of semi-structured surveys and interviews with 16 sport marketing experts. The responses were based on the televised sporting events which the experts managed and/or evaluated. Panel members with the following experience were selected:

- Executives whose primary job was to manage and/or evaluate corporate marketing campaigns through televised sporting events—this status was ensured by examining the relevance of respondents' responsibilities and the relevance of the organization for which they worked.

- Academics specializing in sport marketing—their experience was ensured by examining respondents' books or publications in refereed editions on sport marketing management or evaluation.

In addition, the questionnaire/interview guide ensured that the experts had experience with campaigns of integrated advertising and sponsorship. It was difficult to find the experts, as truly integrated marketing communication is still the exception rather than the rule (Percy et al., 2001). The recruitment stopped after 16 experts were found. This was considered an acceptable number similar to the panel sizes used by other expert-based studies (e.g., Meenaghan, 1994; Richards, 2002).

To warrant safety in numbers and representativeness of the key industry players on the expert panel, the validation was conducted with: 3 academics, 2 advertising firm consultants, 3 corporate marketing/sponsorship/ advertising executives, 2 marketing firm consultants, 2 market research firm consultants, and 4 sport marketing/advertising firm consultants.

The identified management practices were assessed and discussed with experts using a series of self-administered email and fax questionnaires followed by face-to-face and telephone interviews. The questionnaires were designed so that they could be completed and returned both electronically and in writing, and used as discussion guides for the follow-up interviews.

## 3.2 Case Studies

The qualitative case study approach based on semi-structured in-depth interviews was selected to obtain information with minimum bias and to ensure that all the key areas of integration were discussed.

As it was important for the current research to test the constructed theoretical framework by observing the results of multiple cases and obtain comparative data on these cases, a "collective" case study approach was chosen, involving 12 cases.

The first selection criterion was that each of these campaigns had to sponsor a televised event and to place advertisements during the broadcast of this event. Secondly, the campaigns had to promote various types of brands which included sporting apparel, fast-moving goods for consumer and business markets (car petrol and engine oil), consumer durables (cars, watches), consumer services (investment and banking), business and home products (office equipment), and public services. Thirdly, the selected campaigns had to promote the brands through various types of sports—professional and amateur, team and individual, technical (car races) and physical (triathlon), and through various event types—single- and multi-event competition series, national and international.

To select campaigns for the analysis, the lead author videotaped and observed broadcasts of sporting events on Australian television channels during 1998–2000. The events of the selected and analyzed campaigns included the Grand Final of Australian Rules Football, Melbourne, 2000; the Olympic Games 2000, Sydney; and the Wimbledon Tennis Championships, London, 2000. Events containing the brand names or making the brands recognizable are not mentioned here for confidentiality reasons. Some respondents were concerned that information about their companies might be revealed to competitors.

The minimum requirement of each case study was to conduct two interviews—one with an internal respondent from the promoted corporation, and a second with an external respondent from the advertising agency, event management organization, market research firm, or television channel who worked to promote the analyzed corporation. In addition, other professionals involved in the analyzed campaigns were interviewed. Respondents of 12 case studies comprised 12 internal corporate executives (including 5 corporate marketing managers, 4 corporate advertising managers, and 3 corporate sponsorship managers) and 12 external executives (including 8 advertising firm executives, 2 event producers, 1 market research firm executive, and 1 television channel executive). Additional information was obtained through telephone and email discussions with two corporate sponsorship managers and one advertising firm executive.

The 12 case studies were based on observations of broadcasts, interviews with practitioners, secondary data, and internal corporate documents. Discussion guides were emailed or faxed to practitioners prior to the interviews. Respondents emailed or faxed back their initial responses, which were then discussed during the interviews. Therefore, data were obtained from the practitioners' written responses as well as through the interviews. Interviews were tape-recorded where possible and when permitted.

# 4. RESULTS

Table 1 shows how the campaign objectives criteria were rated by the experts and used by practitioners. Sixteen experts indicated the average importance levels of the campaign evaluation criteria on a 4-point scale, and 24 practitioners indicated the number of their campaigns out of 12 which used objectives based on the relative criteria to manage TV advertising and event sponsorship, in integration.

**Table 1.** Relative Importance of Corporate Sport Marketing Campaign Objectives Criteria According to 16 Experts and Campaigns Which Used Objectives Based on These Criteria

| Objectives criteria | Mean importance (16 experts) | Std. deviation (16 experts) | Campaigns (24 practitioners) |
|---|---|---|---|
| Exposure, which occurs when advertisement and sponsorship signage are placed in the media that reaches target audience, measured by size of this audience and by number of promotional messages | 2.9 | 0.8 | 10 campaigns: office equipment, petrol, 5 consumer durables, engine oil, retail chain, government, financial services |
| Processing, or target market's response to the advertisement and sponsorship messages, measured by recognition and recall of the promoted message, media and brand | 3.4 | 0.6 | 9 campaigns: office equipment, petrol, 3 consumer durables, engine oil, retail chain, apparel, financial services |
| Communication effects, measured through attitude and intention to purchase the promoted brand | 3.8 | 0.4 | 8 campaigns: office equipment, petrol, 2 consumer durables, engine oil, retail chain, apparel, financial services |
| Target audience action, sales or market share and profit, which are measures of changed behavior or increased purchases of the promoted brand | 3.3 | 0.9 | 7 campaigns: petrol, 2 consumer durables, engine oil, retail chain, apparel, financial services |

The practitioners' level of objectives use was not consistent with the experts' level of importance of the objectives for integrated advertising and sponsorship:

- 8 of the 12 analyzed campaigns used communication effects objectives indicated by the experts to be the most important (3.8 on a 4-point scale);

- 10 campaigns used exposure objectives, rated to be the least important by the experts (2.9 on a 4-point scale).

The experts indicated in open discussions that it is necessary to utilize all four criteria in sequence, as stated in the following quote:

These four objectives are sort of incremental. You need exposure to have awareness and then you need awareness for attitude change and then attitude change for action. So in that sense all four issues are equally important, in fact they are necessary but not sufficient and only the 4th one is the proof of the pudding but is hard to measure because of confounding variables.

This opinion was consistent with the practitioners' level of objectives use. Consistency was also found between the analyzed campaigns and consumer behavior theory:

- The largest number (10) of the analyzed campaigns set, measured, and achieved the objectives of exposure—the first stage of consumer buying behavior.

- 9 campaigns used objectives of the second buyer behavior stage—processing.

- 8 campaigns set and measured objectives of the third buying stage—communication effects.

- 7 campaigns were able to measure and reach the final buying stage of target audience action.

The experts pointed out that only the most developed campaigns utilized the necessary objectives and measures:

All four criteria are important, but not all companies monitor and comprehensively measure each event and TV show. Measuring impact on the target audience over many months and many events, companies often are not evaluating each particular event or campaign against all important measures. For many companies the primary basis for evaluation is simply the cost of being within a televised event versus the cost of buying standard TV advertising.

The interviews with the 24 practitioners showed that 6 of the 12 analyzed campaigns which had set and measured all four objectives of exposure, processing, communication effects, and audience action, achieved these objectives and increased sales or usage of the promoted brands.

The key objectives of the integrated advertisements and sponsorship promoting a retail chain were to add credibility to the brand through association with the sponsored property and to increase retail sales. Pre-campaign benchmarks were established to measure achievement of the objectives: a set percentage increase on previous year's sales over same period, percentage increase on brand awareness, maintenance (as a minimum) of planned TARP levels, and brand attitude levels. Football suited the researched market well; the objectives and the 'creative' were easy to match to the sponsorship. The campaign achieved its TARPs (exposure), brand awareness (processing), brand attitude (communication effects) and sales (consumer action) objectives. Sales and brand awareness increased strongly over the event period.

The objectives of a campaign promoting a car brand through integrated advertising and sponsorship of multi-stage car races were: to launch a new car model, achieve a low cost per thousand of the target market/s reached, and provide a national, relevant platform for a new product launch. Pre-campaign benchmarks were recorded and objectives were measured against the "hard figures" of cost-per-thousand, reach, frequency, and sales. "Soft" attitudinal survey measures were also used. The results on achieved objectives indicated that the target of sales was exceeded. A new car model was launched into the market effectively and the promoted limited edition vehicles sold out. The reach figure into the target market of the promoted model and the message frequency figure were both high.

In contrast, three analyzed campaigns did not set objectives and evaluate advertising and sponsorship in integration: two campaigns (by a car manufacturer and a governmental organization) accomplished only one of the four objectives (exposure); and a consumer brand campaign did not demonstrate achievement of any goals, confirming the experts' opinion that to achieve the ultimate aim, campaigns should be fully integrated and properly measured.

Sales should be used to track campaign performance. There are not many sport marketing campaigns which compete at this level and which are able to utilize all the integration steps and appropriate methods of campaign evaluation.

Increased sales can only be achieved if a campaign can expose the brand message to target audiences, change the target markets' attitudes toward the brand, and make them want to buy it. One organization of the 12-analyzed aimed not to sell products, but to reduce drunk driving. Such campaigns also require the firm to expose the message to the target audience, change the audience attitude, and convince it to take a desired action. Therefore, the following key objectives of integrated advertising and sponsorship should be set during strategy development and measured after the campaign is launched:

1 To increase exposure which occurs when advertising and sponsorship messages are placed in the media that reach the target audience. Exposure is measured by the size of the audience of television viewers or Target Audience Rating Points, and the number of event spectators and sport followers. The data are obtained through consolidated people meters by research companies such as ACNielsen and through customized material from media monitoring agencies.

2 To enhance processing which is the target markets' response to the advertisement or sponsorship message. Processing is measured by the percentage of target markets who are aware of, and who recall and recognize the promoted message, brand, advertisement, and sponsored event. The data are obtained through customized consumer surveys by market research agencies.

3 To enhance communication effects which are measured through the target markets' desired attitudes, brand image in their minds and intention to take a desired action or purchase the promoted brand. Improved client and employee relations and their perception of brand image are measures specific to sponsorship alone. The data are obtained through customized consumer surveys by market research agencies as well as internal corporate client and employee surveys.

4 To change or increase target audience action, sale, or use of the promoted brand, which is the ultimate bottom-line for integrated advertising and sponsorship. Target audience action, or consumer action, is measured through the percentage of target markets performing the desired action, through product or service sales, market share, or number of customers. The data

are obtained through internal sales system, customer database, as well as through customized surveys by market research agencies. The surveys reveal brand usage and market share through the percentage of consumers indicating use of the promoted brand or service.

Integrated TV advertising and major event sponsorship influence target markets mainly through TV audiences. However, it is also important to use objectives specific to event sponsorship, such as its ability to affect live event spectators, participants, and hospitality guests, including sponsor's business clients and employees. Integrated campaigns may also enable TV commercials to target these audiences, e.g., a financial services company featuring its employees in commercials to improve the corporate brand image among staff, and a sports apparel producer featuring event participants in commercials integrated with the sponsorship of the 2000 Olympics. By doing so, advertisements effectively convey the link between the event and sponsor brands and assist sponsorship to pass the event brand image to the image of the sponsor's brand, ensuring achievement of processing and communication effect objectives.

The results of this study suggested that well-managed and achieving campaigns use their primary corporate goals (e.g., increased sales) within the set of objectives reflecting the process of consumer buying behavior rather than in isolation. The need for systematic and sequential use of buying behavior objectives is consistent with recent recommendations (e.g.,Meenaghan, 2002; S:COMM Research, 2003; Mansourpour, 2007) and is a modification of earlier findings which stressed that corporate sport marketing objectives are highly variable, contingent, and situation-specific (Hirons, 1990;Kuzin & Kutepov, 1994; Quester, 1995).

Different words used for describing the same objectives seem to cause misunderstanding within the industry and create an impression that common sport marketing objectives do not exist.

The objective of processing, or consumer response to the promoted message, was generally described as "awareness" of a brand or advertising or sponsorship, but also was categorized as "advertising and sponsorship recall." Communication effects objectives of improved image or attitude of the promoted brand among the consumers were described as "brand image," and as "consumer attitude."

The results of this study showed that, despite differences in terminology, any objective used by any sport marketing industry personnel to develop, implement, and evaluate corporate marketing campaigns through televised sport can be classified as one of four criteria: exposure, processing, communication effects, and consumer action.

## REFERENCES


1. Arthur, D., Scott, D., Woods, T. and Booker, R. 1998. Sport sponsorship should. A process model for the effective implementation and management of sport sponsorship programmes. *Sport Marketing Quarterly*, 7(4): 49–59.

2. Arthur, D., Woods, T. and Scott, D. Paper presented at the Third International Sport Management Alliance Conference in conjunction with the Fifth Sport Management Association of Australia and New Zealand Conference. Sydney, Australia. *Exploring the sport sponsorship buying centre: an Australian case study*,

3. Belch, G. E. and Belch, M. A. 2009. *Advertising and promotion: An integrated marketing communication perspective, 8th,* Boston: McGraw-Hill/Irwin.

4. Berkes, P. and Nyerges, M. 2004, September. "Business elements in sport: Factors affecting sport sponsorship decision-making process". In *12th European Association of Sport Management European Sport Management Congress Proceedings* 171–174. Ghent, , Belgium CDRom PDF

5. Berkes, P., Nyerges, M. and Váczi, J. 2008. Macro-aspects affecting sport sponsorship: The case of Hungarian professional soccer clubs' sponsors. *Society and Economy,* 29(3): 383–412.

6. Cheng, P. S. T. Paper presented at the Third International Sport Management Alliance Conference in conjunction with the Fifth Sport Management Association of Australia and New Zealand Conference. Sydney, Australia. *The successful sport sponsorship in the new millennium,*

7. Cliffe, S. J. and Motion, J. 2005. Building contemporary brands: A sponsorship-based strategy. *Journal of Business Research,* 58(8): 1068–1077.

8. Collins, L. 2003, January 24–27. Ads aim at Super Bowl home run. *The Weekend Australian Financial Review,* : 54

9. Farrelly, F. 1995. *Marketing practitioners lack the understanding necessary to effectively manage commercial sponsorship—an exploratory investigation,* Melbourne, Australia: Masters thesis. Monash University.

10. Farrelly, F. 2002. *A predictive model of sport sponsorship renewal in Australia,* Adelaide, Australia: Doctoral dissertation. University of Adelaide.

11. Grdovic, S. 1992. *An evaluation of corporate sponsorship*, Melbourne, Australia: Unpublished honors thesis. Monash University.

12. Hal Dean, D. 2002. Associating the corporation with a charitable event through sponsorship: Measuring the effects on corporate community relations. *Journal of Advertising*, 31(4): 77–88.

13. Hansen, F. and Scotwin, L. 1995, August. An experimental enquiry into sponsoring: What effects can be measured?. *Marketing and Research Today*, : 173–181.

14. Hirons, M. 1990. "Sports sponsorship". Melbourne, Australia: Masters thesis. Monash University.

15. Hoek, J. A. 1999. Sponsorship: An evaluation of management assumptions and practices. *Marketing Bulletin*, 10: 1–10. Retrieved October 13, 2008, from http://marketing-bulletin.massey.ac.nz/V10/MB_V10_A1_Hoek.pdf

16. Irwin, R. L., Assimakopoulas, M. K. and Sutton, W. A. 1994. A model of screening sport sponsorship opportunities. *Journal of Promotion Management*, 2(3/4): 53–69.

17. Irwin, R. L and Sutton, W. A. 1994. Sport sponsorship objectives: An analysis to their relative importance for major corporate sponsors. *European Journal for Sport Management*, 1(2): 93–102.

18. Kuzin, V. and Kutepov, M. 1994. *Sport sponsorship*, Moscow: Russian State Academy of Physical Culture Press.

19. Kuzma, J. 1990. *An investigation of the role of events marketing in corporate strategy*, Kent, OH: PhD dissertation. Kent State University.

20. Lawson, M. 2002, December 12. Struggling to put the runs on the board. *Australian Financial Review*,

21. Mansourpour, S. 2007, September. *An investigation into the use of sports sponsorship as a marketing tool*, Cranfield, Bedfordshire, United Kingdom: MSc thesis. Cranfield University, School of Management. Retrieved July 24, 2008, fromhttps://dspace.lib.cranfield.ac.uk/handle/1826/2209

22. McDonald, C. 1991. Sponsorship and the image of the sponsor. *European Journal of Marketing*, 25(11): 31–38.

23. Meenaghan, J. A. 1983. Commercial sponsorship. *European Journal of Marketing*, 17(7): 1–73.

24. Meenaghan, T. 1991. The role of sponsorship in the marketing communication mix. *International Journal of Advertising*, 10(1): 35–47.

25. Meenaghan, T. 1994. *A holistic investigation of image dimensions of commercial sponsorship*, Dublin, Ireland: PhD dissertation. University of Dublin.

26. Meenaghan, T. 1999. Commercial sponsorship—the development of understanding. *International Journal of Sports Marketing and Sponsorship*, 1(1): 19–31.

27. Meenaghan, T. 2001. Understanding sponsorship effects. *Psychology & Marketing*, 18(2): 95–122.

28. Meenaghan, T. 2002, September. *Examining commercial sponsorship*, Presentation to the Expert Group on Commercial Communications.

29. Mellish, M. 2000, October 3. Westpac claims gold in $100 m Olympic outlay. *Australian Financial Review*, : 23

30. Olkkonen, R. 2001. Case study: The network approach to international sport sponsorship arrangement. *Journal of Business & Industrial Marketing*, 16(4): 309–330.

31. O'Reilly, N. J. and Madill, J. J. 2007. Evaluating social marketing elements in sponsorship. *Social Marketing Quarterly*, 13(4): 1–25.

32. O'Reilly, N. J., McCarthy, L., Séguin, B. and Lyberger, M. 2005, May 28–31. "Sponsorship and the Super-Bowl: A longitudinal analysis". In *Proceedings for the 33rd Annual Marketing Division Conference of the Administrative Sciences Association of Canada, Toronto* Edited by: Pons, F. 52–60. Retrieved November 13, 2008, fromhttp://luxor.acadiau.ca/library/ASAC/v26/03/26_03_p052.pdf

33. Papadimitriou, D., Apostolopoulou, A. and Dounis, T. 2008. Event sponsorship as a value creating strategy for brands.*Journal of Product & Brand Management*, 17(4): 212–222.

34. Percy, L., Rossiter, J. and Elliott, R. 2001. *Strategic advertising management*, New York: Oxford University Press.

35. Quester, P. 1995. "Awareness and recall as measures of sponsorship effectiveness: A longitudinal study of the Adelaide Formula One Grand Prix". Adelaide, Australia: Working Paper. University of Adelaide.

36. Quester, P. and Thompson, B. 2001. Advertising and promotion leverage on arts sponsorship effectiveness. *Journal of Advertising Research*, 41(1): 33–47.

37. Richards, J. I. 2002. Oracles on "advertising": Searching for a definition. *Journal of Advertising*, 31(2): 63–78.

38. Rossiter, J. 1999. Measuring the effects of sponsorship. *Research News*, 16(5): 1–4.

39. Rossiter, J. and Percy, L. 1997. *Advertising communications and promotions management*, New York: McGraw-Hill.

40. Lane, R., King, K. and Russell, T. 2008. *Kleppner's advertising procedure, 17th*, Englewood Cliffs, NJ: Prentice-Hall.

41. S:COMM Research. 2003. *Presentation material of the S:COMM research consultancy in sponsorship & sports research and media evaluation*, Chesham Buckinghamshire, , UK: S:COMM Research (UK) Ltd. Copy in personal collection of authors.

42. Shannon, J. R. 1999. Sports marketing: An examination of academic marketing publication. *Journal of Services Marketing*, 13(6): 517–535.

43. Shilbury, D., Quick, S. and Westerbeek, H. 2003. *Strategic sport marketing, 2nd*, Sydney, , Australia: Allen & Urwin.

44. Shimp, T. A. 2000. *Advertising promotion. Integrated marketing communications*, New York: Dryden Press.

45. Shimp, T. A. 2003. *Advertising, promotion and supplemental aspects of integrated marketing communications*, Mason, OH: Thompson South-Western.

46. Smolianov, P. 1994. "An investigation of sport marketing contingency competencies". Provo, UT: Master's thesis. Brigham Young University.

47. Smolianov, P. and Shilbury, D. 2005. Examining integrated advertising and sponsorship in corporate marketing through televised sport. *Sport Marketing Quarterly*, 14(4): 239–250.

48. Spais, G. S. and Filis, G. N. 2006a. The Athens 2004 Olympic Games: An event study. *Journal of Integrated Marketing Communications*: 52–61. Retrieved November 28, 2008, fromhttp://jimc.medill.northwestern.edu/JIMCWebsite/2006/AthensOlympicGames.pdf

49. Spais, G. S. and Filis, G. N. 2006b. Stock market reaction on Olympic sponsorship announcement using event-study method. *Journal of Korean Academy of Marketing Science*, 16(2): 95–108.

50. Stamler, B. 2002, October 29. A cable network offers the ultimate plan to end ad zapping: commercials within programs. *New York Times*, : 8

51. Stensholt, J. 2003, January 23–29. Sponsors enjoy the sports spoils. *BRW*, : 17

52. Stotlar, D. Paper presented at the Fourth Annual Conference of the Sport Management Association of Australia and New Zealand. Gold Coast, Australia. *Sponsorship in North America: A survey of sport executives*,

53. Sylvestre, C. M. and Moutinho, L. 2008. Leveraging associations: The promotion of cultural sponsorships. *Journal of Promotion Management*, 13(3/4): 281–303.

54. Thwaites, D. and Carruthers, A. 1998. Practical applications of sponsorship theory: Empirical evidence from English club rugby. *Journal of Sport Management*, 12(3): 203–219.

55. White, J. and Mazur, L. 1995. *Strategic communications management: Making public relations work*, UK & Reading, MA: Addison-Wesley Publishing. London: Economist Intelligence Unit; Wokingham

# CHAPTER 5

# Develop a flexible method to assess buildings hosting major sports events environmentally through the world

**D. Ahmed Ahmed Fekry[a], Abbas Mohamed El Zafarany[b], Amal Kamal Mohamed Shamseldin[c]**

[a] Faculty of Engineering, Cairo University, Egypt

[b] Faculty of Urban Planning, Cairo University, Egypt

[c] Faculty of Engineering, Ain Shams University, Egypt

## ABSTRACT

Specific international institutions are responsible for managing and organizing major sports events besides choosing the hosting city for those events which is a difficult task, as there is a need for an appropriate decision using highly credible and justifiable mechanisms. Assessing the hosting city includes the assessment of sports buildings used in those events; however the diverse characteristics of countries aiming to host sports events raises the problem of obtaining fair environmental assessment results for the submitted projects. There are already a number of environmental assessment methods of buildings around the world and some were used to evaluate a group of major sports buildings in their countries. A particular version of Building Research Establishment Environmental Assessment Method (BREEAM) was used to assess the sports buildings hosting the Olympic Games in London 2012. However, it cannot be used outside England without the presence of several defects in the evaluation process, especially when dealing with different regional characteristics. Many countries are still without environmental assessment methods of their own, besides the unfairness in the comparison of results from available assessment methods among countries. Difficulty finding a standardized assessment method appears because of the spatial and temporal variables. The paper aims to show the importance of having a flexible method that could adapt to all the variables affecting environmental assessment of buildings with different characteristics and conditions of the countries hosting sports events as well as the different time periods, to get the utmost justifiable and precision results when choosing the hosting city.

## KEYWORDS

Environmental assessment methods of buildings; Organizing major sports events; LEED International; BREEAM International; SBTool

# 1. INTRODUCTION

People around the world, with their different cultures, are looking to sports events with great interest, as a source of excitement, joy, and mutual experiences, besides leading to social communication and cultural cooperation among countries. Various countries are interested in hosting sport events because of its economic, political and social benefits. It is considered a great opportunity for the hosting country and city to upgrade and to present itself to the world. It helps introducing them internationally, leading to a positive impact on business and tourism. Hosting sport events increases the attention to provide appropriate places and infrastructure to host those events, giving a positive impact on citizens as a result of close contact with other communities by gaining social and cultural experiences [1] and [2]. Choosing the hosting city of sport events is not an easy task. The duty of managing and organizing sport events depends on assessing the submitted applications to pick the hosting place carefully. Evaluating the projects is necessary according to specific criteria and guidelines. In the end, choosing the location is strongly related to the ability of the institutions in charge of taking an appropriate decision, in addition to the availability of the facilities according to a clear, credible, fair and highly accurate methodology.

There is a common relationship between the environment and sports, in terms of global attention upon influence. In other words, if sports attract people for fun and excitement, then linking environmental concerns with sports interest reflects the growing environmental awareness across the globe and raises the percentage of the active participation in both activities. When a city hosts international sports events, this will pay back several benefits in exchange for what is spent to host the event, including upgrading and adding value to the environmental constructions which can be monitored and evaluated. The environmental assessment of buildings is considered an integral part of the overall assessment of the urban environment, as it is associated with the effectiveness of those buildings to fulfill the users' needs, in addition to encouraging the demands for sustainable buildings, ensuring the best environmental practice integrated into the buildings. Besides all that, developing standards and levels beyond the ones required by traditional systems and providing innovative solutions would ensure minimizing the environmental impact of buildings on the environment [3].

Cities desiring to host sporting events are not obligated to provide an environmental assessment of their buildings within the introduced project, so they may provide that assessment voluntarily in the way they assume is appropriate, which is often in line with the different methods used in each of them. The above shows the existence of detractions in credibility and fairness of the evaluation results comparison, as well as the neglecting of the environmental assessment role in general, especially with the absence of a unified authority responsible of that assessment and an appropriate mechanism to compare the results. One of the most obvious examples of the weakness of the environmental assessment role when choosing the host cities for major sporting events is what happened in Rio de Janeiro 2012. There is no doubt that the incident which took place in downtown of Rio de Janeiro which is preparing to host the soccer world cup in 2014 and the Olympic games in 2016 increased the importance of the environmental assessment of buildings in the cities hosting such events, as three buildings collapsed in the city center, which sheds light on its aging infrastructure [4].

From the previous, the paper's objective is emphasizing the importance and capability of creating a flexible method that could spread throughout the world, with time to get the utmost justice and precision when choosing the hosting city. This objective can be achieved according to a number of steps. First of all, by determining the current status of the environmental assessment of sports buildings, then determining the problems in comparing the assessed results using one or more environmental assessment methods of buildings, then explaining the challenges facing a fair comparison for these results, hence, determining the current solutions for such challenges. Finally, proposing a solution with a flexible assessment method which can adapt with the impact of spatial and temporal variables to avoid the current defects and to ensure the utmost justice, credibility and transparency of the assessment results.

## 2. ORGANIZING AND MANAGING MAJOR SPORTS EVENTS

International institutions are entrusted to manage and organize different sports events, for example, the International Federation of Football (French: Fédération Internationale de Football Association) (FIFA), is the organizing institution for football around the world, based in Zurich, in Switzerland [1]. The International Olympic Committee (IOC), which is based in Lausanne, in Switzerland also, is the organization responsible for managing and running Olympic Games. Those institutions are handling many tasks, including choosing the location where the sports events are to be held, making sure of the preference of those sites compared to others, and to ensure the validity of the country and city for hosting such events [2].

## 3. ENVIRONMENTAL ASSESSMENT OF BUILDINGS

Environmental assessment methods of buildings emerged across the world to determine the principles and standards that are targeted in the relationship between buildings and their environment. They are used in issuing assessment certificates granted for buildings that confirm their commitment to the environment according to specific classifications that puts them in competition with other buildings. The assessment includes assessing indoor environmental quality, sustainable sites selection and management, water and energy consumption efficiency, materials and resources selection and consumption efficiency, the potential re-use and recycling, besides other criteria which are used to judge the efficiency of the environmental performance of buildings. Building Research Establishment Environmental Assessment Method (BREEAM) is considered the first assessment method, which has been established through the Building Research Establishment (BRE) in the United Kingdom [5]. Several different methods appeared later in different places around the world, such as Leadership in Energy and Environmental Design (LEED) in the United States (1998), which was developed by the American Green Building Council (USGBC) [6]. There is also the Green Star which appeared in Australia in 2003 [7], and the Comprehensive Assessment System for Building Environmental Efficiency (CASBEE) which appeared in Japan in 2004[5] and [8]. The previous four methods are considered to be the most famous and widespread methods in the world. Several versions of the assessment methods were issued to cover different building types, in terms of their functions and age. It is noted that different methods include environmental issues with different weights that represent the environmental importance of these issues according to specialized groups of construction specialists and academics [3], [5] and [9].

A number of countries planned to catch up the first wave of environmental assessment methods by producing their own methods such as Promise E in Finland, Lider A in Portugal, Verde in Spain, and Green Pyramid Rating System (GPRS) in Egypt [10]. Local methods in different countries help governments to encourage and impose environmental compliance of buildings, as well as paying more attention to regional issues that belong to different countries without external influences. They provide the possibility of benefiting from local references such as energy codes, along with the possibility to take advantage of all the previous experiences in the assessment field without depending on specific legislations or institutions. A huge time is consumed to create these new methods, especially when using different resources of expertise, along with the difficulty of putting the local method in competition with the well-known and experienced methods. They also need a lot of time and effort to test them after their release and to use the feedback in developing the method [10] and [11].

## 4. CURRENT STATUS OF THE ENVIRONMENTAL ASSESSMENT OF SPORTS BUILDINGS

It is noted that a number of sports buildings that received environmental performance certificates are located in the producing countries depending on the methods used for their assessment (for example, the American Airlines Arena, Amway Center and Philips Arena, got different classifications of LEED [12], while Edgbaston Cricket Ground, and National Indoor Sports Arena got different classifications of BREEAM). When London won the right to host the 2012 Olympic Games the organizing team undertook the environmental assessment of buildings into consideration, taking into account a detailed assessment that covers a range from the pre-games period to the post-games period. They got a special issuance of BREEAM for these projects, as the Olympic venues need to be evaluated in a different way from the available versions. This evaluation included studying the environmental impacts over the lifetime of buildings not during the games period only [13]. From the above, it is clear that the environmental assessment of sport buildings are linked to existing assessment methods in a limited number of countries that were also the pioneers in the environmental assessment field, which gives them the necessary ingredients and experience in assessment. However, they are also the countries hosting sports events the most [2], so there will be a focus on those countries, reducing the chances of other countries to host such events.

## 4.1 Problems comparing environmental assessment results of buildings hosting sports events

The assessment results of the buildings hosting sports events cannot be compared throughout the world and for different time periods without lacking accuracy and credibility for several reasons as mentioned below.

## 4.2 Problems comparing the environmental assessment results using different methods

The maximum required assessment level of sustainability varies between different countries, as it represents the best environmental practice available that can be accessed in each one, which differs significantly among them. Therefore, the maximum level of assessing items and environmental functions in each method vary according to the local experience and practice in each country. The goals pursued by the different methods to be achieved are variable, even with the common general principles among them [14]. Different practices among countries also lead to the emergence of many differences between these methods, and when exposed to the most famous and popular assessment methods, several differences appear in the assessed issues, weights used for evaluation, classifications, scoring ranges and the legislations used in evaluation. Some variations between assessment methods can be shown in the following table (Table 1) [7], [8], [9], [15] and [16].

Comparative research done by BRE to give approximate values of different methods, namely BREEAM, LEED, Green Star, and CASBEE, revealed that when assessing buildings designed to get a high score in some methods, their evaluation scores may not match their scores in other methods. For example, if a building is designed in the UK to get a high score in LEED, it often gets just a good score with BREEAM [5] (Fig. 1). Another research that compared energy consumption in an office building in Dubai using BREEAM, LEED, and Green Star methods showed that the building which got a high score according to Green Star got a low score in BREEAM and fails to be classified in LEED, since those different methods are using different standards, measurement approaches and rating scales [6]. Environmental requirements in different assessment methods also depend on various assessment criteria, laws and codes used as references in each of them, which clearly can be shown between BREEAM, which is based on European legislation, and LEED, which is based on ASHRAE standards [5]. There is also a difference in the measurement approaches used to evaluate the items' requirements in the different assessment methods. For instance, assessing the efficiency of energy consumption in LEED mainly depends on improving the energy performance according to the percentage of improvement of annual energy cost, while BREEAM depends on the reduction of carbon dioxide emissions. In conclusion, totally different standard buildings are used in the evaluation of the methods in question [6],[9] and [16].

**Table 1.** Some comparative aspects between BREEAM, LEED, green star and CASBEE for multi-residential buildings in the same time period [7], [8], [9], [15] and [16].

| Comparative aspects<br>Producer | BREEAM<br>BRE | LEED<br>USGBC | Green star<br>GBCA | CASBEE<br>JaGBC-JSBC |
|---|---|---|---|---|
| Evaluation areas and points obtained | 2008/upgraded 2010 | 2009/upgraded 2011 | 2009 upgraded 2011 | 2010 |
| | Management (10) | Awareness & education (3) | Building Management (18) | Quality of services (16.6%) |
| | Health and wellbeing (12) | Indoor Environmental Quality (21) | Indoor Environment Quality (20) | Indoor Environment (16.6%) |
| | Energy (23) | Energy & Atmosphere (38) | Energy (26) | Energy (16.6%) |
| | Transport (9) | Location & Linkages (10) | Transport (14) | |
| | Water (8) | Water Efficiency (15) | Water (12) | Resources & Materials (16.6%) |
| | Materials (17) | Materials & Resources (16) | Materials (31) | |
| | Waste (8) | | | |
| | Land use and ecology (10) | Sustainable Sites (22) | Land use & Ecology (11) | Outdoor environment on site (16.6%) |
| | Pollution (12) | | Emissions (18) | Off-site environment (16.6%) |
| | Innovation (10) | Innovation & Design Process (11) | Innovation (5) | |
| Classifications and related grades | Pass (30–45%)<br>Good (45–55%)<br>V good (55–70%)<br>Excellent (70–85%)<br>Outstanding (85–110%) | Certified (45–59 point)<br>Silver (60–74 point)<br>Gold (75–89 point)<br>Platinum (90–136 point) | 4 stars (45–59%)<br>5 stars (60–59%)<br>6 stars (75–105%) | Poor-C (0–0.49)<br>Fairly poor-$B^-$ (0.5–0.99)<br>Good-$B^+$(1–1.49)<br>Very good- A (1.5–1.99)<br>Excellent-S (0.3–) |
| Used legislation | European and UK regulations | American regulations especially ASHRAE | Australian legislation and local protocols | Japanese legislation and codes |

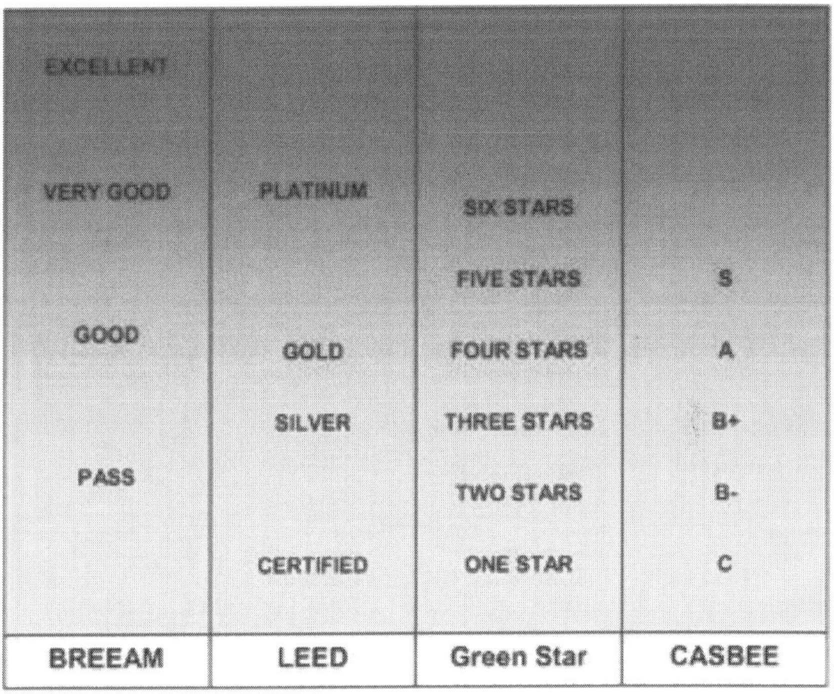


**Figure 1.** Approximate values for different assessment methods used to assess buildings designed in the United Kingdom [5].

There are various issues in assessment among current environmental assessment methods of buildings and their assessing ratios. For example, in LEED there are many items that cover issues not included in BREEAM. I return LEED deals insufficiently with the Ozone Layer depleted materials [5] and [9]. It is noted when comparing CASBEE and BREEAM items, for example, that there are 44 points in the first method that do not exist in the second one, because it is mainly focused on Japan regional issues, particularly in regard to earthquakes [5] and [8]. It is also noted when applying one method like Green Star upon the entire area of Australia covering different climatic zones; some items are not always applicable in some places, like the item of not using cooling towers, which is easily accessible in South Australia [7]. Some variations in the assessed applications appear in some environmental assessment methods too. Although LEED encourages the existence of sufficient parking lots and grants degrees for that, BREEAM for example grants degrees for minimizing those spaces to reduce emissions [3]. Calculating ways used to get the final scores of buildings also vary among different methods, while the most common way is by gathering the grades of each assessment item, as in LEED, BREEAM, Green Star and GPRS. CASBEE and other methods show another way to calculate the final evaluation of the building, which depends on the output of the Building Environmental Efficiency (BEE) [8].

## 4.3 Problems comparing the environmental assessment results using the same method

Current assessment methods aim to achieve the best environmental practice available in buildings. Thereby, the maximum rating of buildings is given for a level of sustainability lower than 100%, meaning that the maximum grade is granted to a building which achieves a proportion of ideal sustainability which may change with time and available constituents [14]. Noted that the previous reason is the one that led to the ratios transfer of the classification rates used in BREEAM when a need to get a higher level of environmental performance appeared over time. So a final rate which is 'outstanding' was added after the rate 'excellent' and was the highest rate until 2008. The remaining classification rates were also modified which led to the emergence of a problem while comparing the building results before and after 2008 [9].

There are other problems which may lead to difficulty in comparing the building results from different versions of the same method, especially when using the final assessment degree in the form of numbers not percentages as in LEED, while assessing residential buildings in LEED was from 69 points in 2005, it became 136 points in 2011. There are also radical changes in the assessing versions for the same method over time resulting in the impossibility of locating the environmental performance of assessed buildings using an earlier version according to a later one, where for example, one area of evaluation may be separated in some versions and emerged in others during different time periods[17] (Table 2). In the issuance of residential buildings of LEED 2011 a factor known as 'Home Size Adjustment' is used to change the final classification assessment ranges of buildings depending on the buildings different sizes. So the rating of a building may start from 35 instead of 45 when the factor = −10, or start from 55 instead of 45 when the factor = +10 [16]. Changing these ranges of rating classification leads to the impossibility of comparing the environmental performance of residential buildings of different sizes with each other. Instead of changing the buildings classification rates it is preferable to include the effect of buildings size on the estimated weights of resources, materials and energy issues which are affected by it, and then change the estimated weights of other issues, so the final classification remains uniform.

Countries hosting major sports events vary; therefore the environmental assessment of buildings hosting such events must be adapted to the spatial variable characteristics of each country. Assessment of buildings dedicated to hosting events according to an American request cannot be the same for buildings dedicated to hosting those events in Egypt for example. Major sports events are held at distant time periods which are enough for the emergence of technological developments or the change of the interests in different environmental issues, therefore assessing the environmental performance of buildings in 2010 cannot be used for 2014, meaning that environmental assessment of buildings hosting sports events must be adapted with time variables too. Thereby, the importance of getting a flexible method for assessing the environmental performance of buildings can be inferred to be able to change and evolve according to spatial and time variables to ensure the fairness of the evaluation results.

Spatial characteristics vary significantly between countries hosting sports events, even between regions in the same country. Natural spatial variables affecting environmental assessment of buildings include climatic, hydrological, geological and ecological characteristics. There are also spatial humanity variables, such as the prevalent practice, culture, prevailing laws, the cost of resources and materials as well as the population density. Different site conditions change the attention to sustainability issues, for example, assessing the efficiency of water consumption in a rainy country differs from those in dry ones. Drought in Australia for example, leads to a rise in the importance of rationalizing the local water, unlike the United Kingdom, where the heavy rain and high population density raise the attention of land use and ecology [6]. Climatic characteristics control the material construction types and determine the used techniques of operating and maintaining buildings [18]. Urban characteristics control the appropriate means of transport and urban spaces. Environmental assessment of buildings is also influenced by the urban and historical characteristics of the country [16].

**Table 2.** Some different aspects between some versions of LEED which lead to the difficulty of comparing their assessment results [16] and [17].

| Some LEED aspects | Some versions of LEED | | |
| --- | --- | --- | --- |
| | Multi-residential 2005 | Multi-residential-2009 | Multi-residential-2011 |
| Evaluation areas and points can be obtained | Sustainable Sites (SS) (14)<br>Water Efficiency (WE) (5)<br>Energy & Atmosphere (EA) (17)<br>Materials & Resources (MR) (13)<br>Indoor Environmental Quality (EQ) (15)<br>Innovation & Design Process (ID) (5) | Sustainable Sites (SS) (26)<br>Water Efficiency (WE) (10)<br>Energy & Atmosphere (EA) (35)<br>Materials & Resources (MR) (14)<br>Indoor Environmental Quality (EQ) (15)<br>Innovation & Design Process (ID) (6)<br>Regional Priority (RP) (4) | Innovation & Design Process (11)<br>Location & Linkages (LL) (10)<br>Sustainable Sites (SS) (22)<br>Water Efficiency (WE) (15)<br>Energy & Atmosphere (EA) (38)<br>Materials and Resources (MR) (16)<br>Indoor Environmental Quality (EQ) (21)<br>Awareness & Education (AE) (3) |
| Classifications and related grades | Certified (26–32 point)<br>Silver (33–38 point)<br>Gold (39–51 point)<br>Platinum (52–69 point) | Certified (40–49 point)<br>Silver (50–59 point)<br>Gold (60–79 point)<br>Platinum (89–110 point) | Certified (45–59 point)<br>Silver (60–74 point)<br>Gold (75–89 point)<br>Platinum (90–136 point) |

## 4.4 Challenges with environmental assessment of buildings hosting major sports events

Other variables connected to time appear when organizing sport events, especially the change of global environmental issues interest priorities over time, including the issue of global warming, scarcity of fresh water resources, degradation of biodiversity and others. There are also variables associated with the degree of technological development, which is connected to the different elements and components used in buildings. New materials or inventions may appear, importance of some resources may be discovered and other forms of transformation in events and ideas may arise, resulting from researches or studies leading to change in vision of many assessment items. Since the World Cup is held every four years and the Olympic Games are organized currently every 2 years with alternating summer and winter games, those time periods are sufficient for the emergence of time effect on the environmental assessment components of the buildings.

## 4.4 Current solutions of variables' impact on buildings assessment results

International versions of environmental assessment methods of buildings such as BREEAM and LEED known as BREEAM International and LEED International are adjustable versions by the green buildings' councils scattered around the world. They can be converted into a local method for each country, helping to form a locally recognized assessment tool, and keep the advantage of benefiting from the well-known methods experience [9] and [19]. The idea of international versions is unifying a set of constants with the original method and letting the region teams complete them. So the areas and essential assessment items are similar with the original method, while the weights of relative importance of the assessed fields and items and the number of items within the assessed fields as a whole are different. Changes made in these versions are required to be as minimum as possible [18] and [19]. However, some defects have emerged in the dealing of those versions with variables affecting the environmental assessment of buildings, and some of them can be displayed as follows:

- Amendment consumes a lot of time and effort because of its association with more than one aspect, besides the requirement of the approval of the original method's institution for the amendment requests made by the local authorities, in order to preserve the consistency and to protect the brand promotion of the original method[19].

- Standards used in assessment are not to be changed unless the substituted standards are equally strict or stricter than the existing standards, which cannot be achieved in many local codes around the world, wasting the chance of taking advantage of them [19].

- International versions are affected by the view and culture of the original producing country. They are also affected by their market's requirements and they continue to deal with the assessment items requirements in the same manner of the original methods [3], which can be noted clearly in the significant differences of assessed issues and weights between BREEAM Gulf and LEED Emirates which are designed for the same region [3] and [19].

- International versions depend on deleting the specialized items and keeping general ones, which are commensurate with all buildings and different places. However, that way leads to empty the method from experiences which were included in it. This may cause work duplication, waste of time and effort, and experiences conflict when putting items that existed previously [11].

Green Building Challenge (GBC) appeared in Canada to deal with variables affecting the environmental building assessment. It was under construction since 1996 through a variety of specialists and was handed to the International Initiative for a Sustainable Built Environment (iiSBE) in 2002. GBC was created to help countries producing their own assessment tools. In the Sustainable Building Conference 2002 (SB 2002) in Oslo, Norway; Green Building Tool (GBTool) was introduced as an assessment tool for GBC which was upgraded later to Sustainable Building Tool (SBTool) [14] and [20]. The idea of this method depends on developing general values that can be replaced with local values by local experts to determine the levels of appropriate performance [20]. GBC has no limits in the amount and type of changes which are to be made in it as long as they are necessary. The items' estimated weights are edited in the method by using a scale of 1–3 to express properties of items in terms of the extent, intensity and duration of potential effect of each assessed item. When analyzing the dealing of GBC with variables a number of defects could be displayed as follows:

- Variables' impacts are gathered in common and similar characteristics for all items, while instead of specifying certain properties to include the effects of different variables, it is preferable to study the effect of each variable according to its respective characteristics [20].

- The used technique does not allow distinguishing the effect of some variables from others for various items, as including the impact of variables through specific characteristics does not separate the effect of one variable from another variable for different items, so it is difficult to distinguish how much effect each one has when determining the value and weight of items [14] and [20].

- The maximum assessment level does not represent achieving the perfect sustainability for the item's requirements, which makes the same level to be used in expressing different levels of sustainability in different countries and over time, which detracts from the acquired feature achieved by using several levels to assess the achievement of items' requirements, ranging from negative practice (−2) to best practice (+5) [14] and [20].

- There is no total environmental performance results for assessed buildings that can be compared with other buildings' results, as the assessment process is divided into different stages without having an appropriate mechanism for combining different assessment results of these stages into one. This makes it difficult to compare the environmental performance of assessed buildings except for the same stage, especially for GBC method with the possibility of deleting some of the assessed main issues in some of those stages depending on variable effects for different countries [14] and [20].

## 4.5 Proposed solution of variables' impact on sports buildings' assessment results

Instead of giving the task of the environmental assessment of the buildings of major sports events to the countries applying to host these events, the formation of a competent organization for the environmental assessment of these buildings is preferred, which can be based in Switzerland, for example, like the other major organizations of those events. The proposed organization may include environmental assessment experts from all around the world, who are experts in this field and independent of other assessment organizations. Those experts configure different versions according to the local environmental conditions of the applicant countries for hosting, taking into consideration the time period to ensure the utmost justice, credibility and transparency of the assessment results, and to ensure a fair assessment and a consistent environmental building performance from the perspective of green architecture for all countries. The challenge in this case is to facilitate the work of experts by providing a flexible assessment method that can adapt with the impact of spatial and temporal variables.

## 4.6 Importance of including variables' effect on the assessment results of sports buildings

Due to the great diversity in the characteristics of countries hosting major sports events, there is a need to include the effect of spatial and temporal variables on the environmental assessment of buildings results to get them in the utmost justice and accuracy. This leads to the exclusion of relying on one of the most famous assessment methods such as LEED and BREEAM as a result of their local properties, besides excluding the use of international versions, as there are problems in using them as previously mentioned and also the use of SBTool to create different assessment methods for each country, as there are other shortcomings. The solution of having a local method for each applicant country to host sports events to be used for the environmental assessment of their buildings is not sufficient too, as it is noted that many countries are still without environmental assessment methods of their own and some countries rely on other countries' assessment methods. Even when assuming the existence of local environmental assessment methods for all the countries applying to host sports events, a problem in the credibility of comparing their results appears due to the big differences between their components, weights, rating scales, and the degree of strictness in the used standards, laws and codes.

Even when overlooking the previous differences, another problem appears that forces each country to put a specialized version for the major sports buildings, which leads to a loss of time and effort for the applicant countries which will not be accepted for hosting the event. It is also noted that the characteristics of these buildings are unique and cannot be applied to other buildings later on, which means that they also cannot be reused efficiently by time, as they have to be updated every time to keep up with the time variables affecting the evaluation.

## 4.7 Appling a proposed method for solving variables' effect on the buildings assessment results

It is suggested that a group of experts would identify spatial and temporal variables affecting the environmental building assessment using reliable electronic global sources, then linking them with the method's components and elements affected by them. Variables' properties can be determined by their impact on the environmental assessment of buildings included as follows:

- Experts responsible for determining the presence and the mandatory degree of items study the effect of different variables on each item to identify the variables that eliminate some items or increase their attention to the degree of compulsion.
- Experts responsible for changing the formulation of assessment items determine the impact of spatial and temporal variables on the requirements presented in the formulation of the assessment items. This may include numbers, ratios and required properties to be achieved, besides standards, codes, and laws to be

followed. Experts would then change previous requirements with suitable ones according to the impact of variables.

- Initial weights are put into the different assessment fields and items either by distributing the overall percentage of 100% evenly on the assessment fields then on the items consisting of each detailed level, or by using initial default weights. Experts responsible for modifying weights of assessment fields and items study the impact of different variables in terms of their effect that may be raising or lowering the weight. They would also determine the degree of impact and the degree of variable importance in relation to other variables affecting the same item. Finally, they can use mathematical equations to obtain degrees that can be added or subtracted from the initial weight of each item to get its final degree after modification, noting that changing the weight of any item leads to a change of the weights of all other items that are in the same level, to get a total of 100% of the overall weight of the building.

- Modified weights are examined to ensure the possibility of the building to succeed according to the required limits of success –if found– after completing their modification. It is suggested that the success requirements should include a minimum level of achieving various environmental functions in the building.

- Experts responsible for determining the scoring levels of different items study the impact of different variables upon each item, allowing the identification of scoring levels that commensurate with each item's requirements contained therein. The minimum scoring level may represent a 0% and the maximum is always 100% expressing the ideal sustainability for achieving all items. Experts then determine grades corresponding to each previous level ranging from 0 to 1 and depending on the number of those levels.

- The assessment method appears finally in the form of a set of choices for each of the items, according to what the experts have determined for the scoring levels, depending on the reformulated requirements of each. Then the degree of chosen level during assessment is multiplied by the item's weight already determined by experts too.

## 4.8 Illustrative example of including variables' effect on items in the proposed method

The following example will use one of the environmental assessment building issues to show the main concept of including the variables' effect on items way in the proposed method, hence, showing the possible flexibility through these items. The assessing issue (Energy in Use) is the one chosen to be used in the example. Some main items of this issue are:

Item 1: All buildings shall be designed to minimize carbon emissions and energy demand to achieve a minimum 15% improvement over Part L 2006.[1]

Item 2: Efficiency conversion and distribution of energy obtained through a Combined Cooling, Heating and Power (CCHP) system to provide a minimum 20% carbon reduction.

Item 3: Across the site as a whole, sufficient On-site renewable energy generation capacity shall be installed to meet at least 20% of the annual carbon emissions (reduction) of the venues, and other buildings to be retained within the site in the Legacy phase.

Item 4: Use reasonable endeavors and subject to obtain requisite consents, to seek achievement of a reduction in carbon emissions (against 2006 Building Regulation standards) for the built aspects of the Development of 50% by 2013 [21] and [22].

To change the previous items into flexible ones by including the variables' effect on them the following steps may be used:

First: Depending on the spatial and temporal characteristics of the project, those can be known from some reliable global websites, experts responsible for determining the presence and the mandatory degree of items decide if any of these items is neglected or mandatory. For example all of these items may be presented in an amended method, and (item 1) may be mandatory as it expresses the minimum level of achieving energy efficiency (Table 3).

**Table 3.** Simplified example of determining the presence and the mandatory number of items according to variables' effect.

| Item no. | Presence | Mandatory |
|---|---|---|
| Item 1 | Y | M |
| Item 2 | Y | – |
| Item 3 | Y | – |
| Item 4 | Y | – |

Second: Depending on the spatial and temporal characteristics of the project; experts responsible for changing the formulation of the assessment items change the items requirements according to deferent and new conditions; the requirements of the previous items appear underlined, these underlined sections of the items are the ones which are changed in the formation of the items. In Egypt for example one of the previous items can be:

Item 1: All buildings shall be designed to minimize carbon emissions and energy demand to achieve a minimum 10% improvement over the Egyptian Code to improve the efficiency of energy consumption (if there is no Egyptian code for sport buildings it could be changed to a global one or another country's code that is near to the Egyptian characteristics).

Third: Initial weights are given for the previous items, for example they may all be assessed from 2% as initial weights (taking into account that the overall percentage of assessing the building is 100% including the other items for all other issues), then depending on the spatial and temporal characteristics of the project experts responsible for modifying weights change the initial weights according to these characteristics. For example one of the items weights could be changed according to different variables as follows: (Table 4)

Forth: Depending on the spatial and temporal characteristics of the project; experts responsible for modifying weights put required limits of the building success. The required limits to succeed could be a minimum level of achieving various environmental functions in the building, for example the minimum level of achieving all the environmental functions included in the assessing method (including the Energy in Use issue) may be 7.5%, so experts when modifying items weights they should be careful not to be less than 7.5% for the overall weight of each environmental functions and not more than (100% – (7.5% * no. of environmental functions)), therefore experts after modifying items weights make sure of the possibility of the building to succeed according to the modified weights.

Fifth: Depending on the spatial and temporal characteristics of the project; experts responsible for determining the scoring levels of different items change their scoring levels according to the type of these items and their requirements, and according to the variables affecting them, noting that the maximum scoring level should always reach the 100% of sustainability for each item, for example the first item can have a scoring level as follows: (Table 5)

**Table 4.** Simplified example of determining the weight of an item according to some variables that could affect it.

| Item no. | Initial weight | Variables affecting the item's weight (e.g.) | Type of impact | Degree of impact (DI)[a] | Degree of variable importance (DVI)[b] | Final weight[c] |
|---|---|---|---|---|---|---|
| Item 1 | 2% | V1 (climate) | + | 10 | 5 | 13.5% [d] |
| | | V2 (surrounding urban) | − | 4 | 6 | |
| | | V3 (historic fetchers) | − | 3 | 2 | |
| | | V4 (population density) | + | 4 | 1 | |

[a] (DI) is a number that experts put from 1 to 10 to express the amount of variable impact on the item's weight, number 1 expresses the lowest impact.

[b] (DVI) is a number that experts put from 1 to 10 to express the degree of variable importance in relation to other variables affecting the same item; number 1 expresses the highest importance.

[c] The mathematical equation used to get the final weight is: Final weight = Initial weight + ((DI for V1/DVI for V1) × type of impact × 3%) + ((DI for V2/DVI for V2) × type of impact × 3%) + ...etc. Noting that the figure 3% (proposed and not constant) used in the equation is expressing the even division of 100% (the overall score of the building) on the expected final overall items contained in the assessing method, this figure keep the changing in the item's weight in an appropriate range, So, for the previous proposed figures the final weight was calculated as follows: Final weight = 2% + {((10/5) * 3%) + (−(4/6) * 3%) + (−(3/2) * 3%) + ((4/1) * 3%)} = 2% + 6% − 2% − 4.5% + 12% = 13.5%.

[d] It should be noted that raising or lowering any item's weight will affect the weights of all other items that are in the same level, to get a total of 100% of the overall weight of the building, and it should be noted that the upper levels are calculated before the lower levels, so changing items weight and their initial weights are limited within the weights of the upper levels.

**Table 5.** Simplified example of determining the scoring levels of an item according to some variables that could affect it.

| Item's requirement | Level of achievement | Score of achievement | Item's weight (%) | Final score (%) |
|---|---|---|---|---|
| All buildings shall be designed to benefit any carbon emissions | 100% | 1 | 2 | 2 |
| All buildings shall be designed to block any carbon emissions | 90–100% | 0.95 | | 1.9 |
| All buildings shall be designed to minimize carbon emissions and energy demand to achieve a minimum *80%* improvement over ("*A more strictness regulation than Part L 2006*") | 80–90% | 0.9 | | 1.8 |
| All buildings shall be designed to minimize carbon emissions and energy demand to achieve a minimum *80%* improvement over (*Part L 2006*) | 70–80% | 0.85 | | 1.7 |
| All ... a minimum *65%* improvement over (*Part L 2006*) | 60–70% | 0.8 | | 1.6 |
| All ... a minimum *50%* improvement over (*Part L 2006*) | 50–60% | 0.75 | | 1.5 |
| All ... a minimum *35%* improvement over (*Part L 2006*) | 40–50% | 0.7 | | 1.4 |
| All ... a minimum *15%* improvement over (*Part L 2006*) | 30–40% | 0.65 | | 1.3 |

Experts could also make a mix of scores when needed to express achieving levels of requirements in a way that are not previously stated, for example, in the previous item's requirements achieving 80% improvement over "A more strictness regulation than Part L 2006" leads to get a score of 1.8%, while achieving 15% improvement over Part L 2006 leads to get a score of 1.3%, so achieving 15% improvement over A more strictness regulation than Part L 2006 may lead to get a score of 1.5%.

## 4.9 Some comparison aspects between the proposed method and other current methods

The following shows some comparison aspects between the proposed method and other current methods, which were previously mentioned in the research paper and considered as the current solutions to include variables' effect on buildings assessment results, which is the main concern of the paper (Table 6).

## 4.10 Benefits of implementing the proposed method to buildings hosting sports events

Application of the proposed flexible method helps to achieve a set of objectives that can be summarized as follows:

**Table 6.** Some comparison aspects between the proposed method and other current methods.

| Aspects | Assessment method | Comparisons |
|---|---|---|
| Flexibility in modifying the methods components | International versions (BREEAM International – LEED International) | Modification and changes in these methods are restricted with the lowest possible changes, to unify the main form of the method and the main elements all over the world with the original ones, which reduces their flexibility |
| | GBC (SBTool) | Modification can happen in all its components and elements as long as is necessary to match with the country conditions, but it also allows changing the main presence of the assessed main issues, which leads to a difficulty in comparing the resulting versions between the different countries |
| | Flexible method | Modification and changes are adjustable without the compliance with a final form; although it maintains the unification of the main assessment issues between all the countries to help comparing their results |
| Flexibility in modifying the assessing weights | International versions | Include a minimum degree of success in the different assessed issues beside a set of mandatory requirements, but those limits vary between the countries, and there are no minimum limits for them to prevent them being too low in some cases, and there is no relationship between those limits for the same building |
| | GBC | Does not include limits of success connected to the building |
| | Flexible method | Includes a minimum degree of success, such as a minimum degree to pass achieving the environmental functions included in the method, so changing the assessment items weights does not lead to overcoming these limits which are all connected with each other for the same building |
| Flexibility in dealing with different standards | International versions | There is a strict compliance with the stringent degree of the used standards regardless of its preference in dealing with the local characteristics of different countries |
| | GBC | Allows flexibility to deal with different standards but their stringent degree had no effect on the degree of evaluation |
| | Flexible method | Allows the flexibility to deal with different standards such as the country-specific standards or the global standards or a combination when needed, taking into account their stringent degree which is itself a variable affecting raising or lowering the weights and score levels of assessment items that use those standards |
| Flexibility in modifying the rating scores | International versions | Allow changing the predefined rating scores in the original method without the possibility of having other evaluation levels or secondary ones, noticing that the maximum score for evaluating the items may be given to a degree of sustainability less than 100% |
| | GBC | Allow changing the rating scores of several evaluation levels, noticing that the maximum score for the items are given to a degree of sustainability less than 100% |
| | Flexible method | Allow flexibility of having several evaluation levels, taking into account that the maximum degree is always given when achieving a 100% sustainable level |
| Responsible party for forming the method | International versions | They lead for losing a lot of time and effort because of their association with more than one association working on the amendment of the methods |
| | GBC | Each country can configure their local assessment method on their own without putting a unified level of experts' experience formatting these methods |
| | Flexible method | Depends on a specific and competent organization which includes experts from all over the world according to an appropriate level of experience, it may also engage local experts when needed to avoid the effect of a certain view or culture on the produced method |

## 4.11 Justice in comparing assessment results across time and place

A flexible method, due to its ability to adapt to changes affecting the assessment, helps to get an appropriate accuracy and fairness for the results to assess the environmental performance of buildings among different conditions and country characteristics hosting the sports events and among various time periods.

## 4.12 Providing a global design standard for sports projects

A designer can use versions of the flexible method to make sure of the environmental performance of his building, as buildings hosting such events have unique properties. Thus, the existence of environmental assessment certificates from a specialized authority helps putting global environmental standards to assess the design of these buildings and to meet the environmental requirements that must be taken into account for projects that receive major sports events commensurate, with different characteristics of countries and time periods.

## 4.13 Ability to add different extra items to assess the distinction of sports buildings

The proposed flexible method helps including a range of items that assess the distinction of sports buildings according to the variables associated with each country, where it can assess the social benefits, upgrading of the historical places, elevation of the environmental awareness, as well as the emotional interaction in the sports buildings. It is noted that all of these requirements may vary among different countries.

## 5. LIMITATIONS OF IMPLEMENTING THE PROPOSED METHOD

Applications of the proposed flexible method include some limitations that can be summarized as follows:

Needs an unknown period of time to be examined and edited before competing with other international methods.

- Faces some challenges in marketing its publications on the contrary of other international methods due to the strength of their global institutions.

- Includes several steps to ensure fair comparison of assessed results, which are reflected on both the experts responsible for the method formation and the assessors responsible for the building assessment results, so it is theoretically more difficult to be dealt with than the previous ones and more time-consuming.

- Needs a formation of a unified competent organization for the environmental assessment of buildings to create the different versions of the method in study, while the formation of such an organization needs an unknown period of time, beside the difficulty of gathering appropriate experts from different countries around the world within an appropriate time.

However, previous problems can be resolved. To achieve rapid and accurate versions from the flexible method an automated tool would be recommended and used, as it will facilitate the work of experts and assessors. The tool could be linked with several internet websites, which will help in its marketing and in competing with other international methods, besides helping in the communication with appropriate experts across the network until the formation of a unified organization.

## 6. RESULTS

- Environmental assessment of buildings holding major sports events contributes to the choice of the hosting city for these events according to global environmental standards and bases, creating a competition among them and encourages the achievement of the environmental dimension in buildings that attract millions of people.

- There is a range of spatial variables between applicant countries to host sports events and temporal variables between periods of their occurrence. That must be taken into account when assessing buildings prepared for these events environmentally to ensure fairness of the evaluation results.

- Differences between methods put limits to compare the results of environmental assessment of buildings in a fair and accurate manner among different countries.

- Some current solutions of issuing environmental assessment methods internationally can be used to transmit methods all over the world, as using the international versions of some well-known methods (LEED-BREEAM) or using the SBTool. But despite them dealing with some problems associated with comparing assessment results across places, they possess deficiencies in their way of implying the impact of variables which causes a reduction of the accuracy and fairness of comparable results.

- The high experience of a limited number of countries in the field of environmental assessment of buildings raises their opportunity to assess the hosting sports events buildings. But since these countries also host most of those events, this reduces the chances of other countries to host such events when taking the environmental assessment into consideration.

- It is a difficult request to depend on the applicant countries hosting sports events to do their own environmental assessment for their buildings, especially when taking into account the time and effort to produce specialized versions of their local methods to assess such buildings, which may not be utilized at all or later, besides the lack of credibility in comparing the outcoming results.

- A flexible environmental assessment method of buildings can be used to include the impact of spatial and temporal variables by determining the effect of different variables on each item and then studying their impact on the formulation of those items, weights and evaluation scoring levels.

- The application of the proposed flexible approach to assess the buildings hosting major sports events helps to ensure the fairness of comparing the assessment results across place and time and provides global standards for designing sports projects, as well as the ability of adding various additional items to distinguish those buildings.

## 7. RECOMMENDATIONS

- The competent authorities of organizing and managing major sport events are recommended to configure a neutral subsidiary to issue certificates for environmental assessment of buildings for those events, to evaluate the submitted projects of hosting from different countries and for different time periods.

- The proposed authority for environmental assessment of buildings hosting sport events is recommended to use a standardized assessment method, which should be flexible enough to accommodate to changes in time and place.

- Green councils and competent authorities are recommended to develop and propose a flexible method of assessing the premises of major sports events.

- Green Building Council in Egypt is recommended to gather the Egyptian different variables to study their reflection on their own method or on the produced versions from the flexible method in the future for any suggested place to host a sports event.

## REFERENCES

1. FIFA World Cup. 2012. Available from:<http:// ar.fifa.com/ aboutfifa/organisation/index.html>.

2. The Olympic Movement, International Olympic Committee (IOC). 2012. Available from:<http://www.olympic.org/ioc>.

3. Building Research Establishment (BRE), BREEAM-BRE Environmental & Sustainability Standard- BES 5063: Issue 2.0, BREEAM Gulf 2008 Assessor Manual. BRE Global, 2008.

4. Reuters. 2012. Available from:<http:// ara.reuters.com/article/ worldNews/ idARACAE80P0FW20120126>.

5. T. Saunders. A Discussion Document Comparing International Environmental Assessment Methods For Buildings, 2008.

6. Y. Roderick, D. McEwan, C. Wheatley, C. Alonso, Comparison of Energy Performance Assessment between LEED, BREEAM and Green Star. Eleventh International Building Performance Simulation Association (IBPSA) Conference, Integrated Environmental Solutions Limited, Glasgow, Scotland, UK, 2009.

7. Green Building Council Australia (GBCA), Green Star – Multi Unit Residential. V1, 2009.

8. Institute for Building Environment and Energy Conservation (IBEC), Japan Green Build Council (JaGBC), Japan Sustainable Building Consortium. CASBEE-NCe_v.3.1, 2008.

9. Building Research Establishment (BRE), BRE Environmental & Sustainability Standard- BES 5064: Issue 1.0, BREEAM Multi-Residential 2008 Assessor Manual. BRE Global, 2008.

10. The Arab Republic of Egypt-Ministry of Housing, Utilities & Urban Development through the Housing and Building National Research Center (HBRC). The Green Pyramid Rating System (GPRS). First Edition, National Housing & Building Research Center, 2011.

11. S. Dirlich. A Comparison of Assessment and Certification Schemes for Sustainable Building and Suggestions for an International Standard System. TU Bergakademie Freiberg, The IMRE Journal 5 (2011).

12. The Natural Resources Defense Council (NRDC). Greener Building with LEED (2012). Available from: <http://mlb.greensports.org/greener-building/leed/>.

13. Building Research Establishment (BRE) Global. BREEAM and London 2012 -Development of a Bespoke BREEAM for the

Olympic Venues (2012). Available from: <http://www.breeam.org/page.jsp?id=436>.

14. International Initiative for a Sustainable Built Environment (IISBE). Procedures for Using SBTool 2007, 2008.

15. Building Research Establishment (BRE) Global. BREEAM Multi-residential 2008-Scheme Document SD 5064, 2010.

16. U.S. Green Building Council (USGBC), LEED for Homes Mid-Rise Project Summary, 2011.

17. U.S. Green Building Council (USGBC), LEED for New Construction& Major Renovations, 2005.

18. Aubree. BREEAM International, BRE Global, 2009.

19. Arabian Business Staff Writer. Emirates LEED System Launched, 2007.

20. N. Larsson Overview of the SBTool Assessment Framework. The International Initiative for a Sustainable Built Environment (IISBE) and Manuel Macias UPM, Spain (2012)

21. D. Epstein, A. Young, H. Knight, J. Carris. Learning Legacy – Lessons Learned from the London 2012 Games Construction Project. Olympic Delivery Authority (ODA), London Organizing Committee of the Olympic Games and Paralympic Games Limited (LOCOG), London, 2011. Available from: <http://learninglegacy.independent.gov.uk/index.php>.

22. Olympic Delivery Authority (ODA), London Organizing Committee of the Olympic Games and Paralympic Games Limited (LOCOG). The Environmental Sustainability of the London 2012 Olympic and Paralympic Games, London, 2011.

# CHAPTER 6

# Spatial scale, time and process in mega-events: The complexity of host community perspectives on neighbourhood change

**Julie Clark[a], Ade Kearns[a], Claire Cleland[b]**

[a] Urban Studies, School of Social and Political Sciences, University of Glasgow, United Kingdom

[b] MRC/CSO Social and Public Health Sciences Unit, University of Glasgow, United Kingdom

## HIGHLIGHTS

- Host communities experience a mega-event as a process with strong emotional impacts.
- Perspectives were varied and nuanced, rather than binary 'for or against'.
- Participant views varied between pre-event, during and post-event time phases.

- New housing was welcomed but there are concerns about future management.
- The analysis highlights concerns about market-based processes of change.


## ABSTRACT

A focus on the 'mega' aspect of hallmark events can divert attention from the micro – those local communities who are most impacted by the event. Similarly, attention to the 'event' aspect underplays the long process of bidding and preparation before any putative legacy of urban transformation for local people. This paper uses qualitative data to unpack the complex and multi-layered views of local residents, living in a deprived neighbourhood beside the Glasgow 2014 Commonwealth Games site in Scotland. They reflect on five years of intensive urban regeneration, evaluate the experience of 'lockdown' at Games time, and consider their hopes and fears for the future of the community. Interviewing a mixture of lifelong, established, new and returning residents, we found considerable common ground across the different groups in terms of hopes for a new, mixed community in the area. However, findings also highlight concerns around urban governance practices and the limitations of a market-led approach to regeneration.



## KEYWORDS

Mega-events; Regeneration; Community; Gentrification; Housing

# 1. INTRODUCTION

## 1.1. Mega-events and the micro-geography of impacts

Mega-events can be considered an integral component of much 20th century urban development (Muñoz, 2006), with urban transformation and 'legacy benefits' used to justify the expenditure (Essex and Chalkley, 1998, Leopkey and Parent, 2011, Pound, 2003 and Smith, 2012). In many ways, the essence of a mega-event is scale. Müller (2015) argues that there are four integral dimensions, along which scale should be considered: visitor attractiveness; mediated reach; cost; and transformative impact. Hosting a mega-event has been described as 'one of the most fundamentally political acts of the modern age' (Horne and Whannel, 2012: 204) which, necessarily, advantages some and disadvantages others. Indeed, securing and delivering a mega-event speaks to the power of the host city's elite (Liao and Pitts, 2006).

As Müller's framework demonstrates, the criteria that define a mega-event predominantly involve macro-scale interests. Reflecting this, evaluation data and impact assessments generally use aggregate monitoring data at city, regional or national levels eliding uneven development and differential community-level experiences (Kasimati, 2003, Preuss, 2004 and Owen, 2005). Furthermore, well-intentioned, elite attitudes towards the preferences of relatively disadvantaged groups may be based on assumptions rather than knowledge (Ahlfeldt et al., 2012).

However, this paper prioritises a neighbourhood 'host community' perspective on the impacts of a mega-event, using the Glasgow 2014 Commonwealth Games as a case study. The 'transformative impact' for the built environment and population (Müller, 2015) draws attention to specific geographies, with stadium-building, new 'village' accommodation and adapted transport systems having differentiated effects across the city. It is in the host neighbourhood of a mega-event that the most dramatic impacts can be seen and where, despite a promised legacy, the interests of those promoting and managing the event do not necessarily align with those of those most affected by it. Although urban regeneration and (positive) legacy have increasingly been used as a rationale for undertaking large sporting or cultural events, the question of 'who benefits?' is still germane (Coaffee, 2012, Essex and Chalkley, 1998 and Hall, 1992).

This paper argues that neighbourhood-level geography and time are as salient as 'mega' scale in understanding the nature and impacts of a mega-event. We begin by examining the grounds on which the validity of event-led regeneration has been challenged from the host community perspective. Following this, the Glasgow 2014 Games (GCWG), the study area and research sample are introduced. An exploration of local residents' perceptions of the nature and drivers of change in their neighbourhood, their attitudes towards GCWG regeneration, and their fears and ambitions for the future is used to illuminate the case. The paper closes with reflections on the limitations of mega-events as a catalyst for regeneration and the scope for sustainable, mixed community development around the regeneration site in the post-event era.

## 2. HOST COMMUNITY IMPACTS: THE MEGA-EVENT AS PROCESS

It is tempting to think of the mega-event as a point in time. However, for the host community, from bid, to event, to aftermath, it is more appropriately considered as a process. Physical space is reshaped in order to accommodate the event and social relations are reconfigured, influencing how local people and place are understood, by themselves and by others (Gaffney 2010). From this perspective, time, as well as geography, is a crucial factor in considering the community impacts of a mega-event. We consider the existing literature around three phases, with the host community seen as disadvantaged in each case.

## 2.1. Pre-event: physical displacement and development

The pre-event phase comprises decision-making about the event and associated regeneration, with displacement and demolitions preceding the construction of new infrastructure. Here, the host community often finds itself in opposition to planned developments, being seen as an entity to be managed or maligned in order to facilitate the delivery process.

Despite the high profile of legacy, the language of policy planning can serve to obscure both agency and who benefits (Marcuse, 2015). Although 'the city' is identified as bidding for and hosting a mega-event, in the triumvirate of the state, community and capital, it is the community that has the weakest voice. High economic stakes and fixed deadlines systematically militate against successful democratic participation, and elite actors rather than local communities drive the urban development agenda (Garcia, 2004, Hayes and Horne, 2011 and Surborg et al., 2008).

From the organisers' perspective, community involvement constitutes a 'risk factor', which can interfere with project delivery, while resistance to demolitions or development is construed as not 'business-friendly', and often suppressed or downplayed (Raco 2014: 183; 194; Lenskyj, 2002). Fussey and colleagues note a pattern of public engagements taking place in a limited form and 'only after key decisions have been taken' (2011: 238). Community actors hoping to influence events require the capacity to identify and engage assertively with layers of complex and changing bureaucratic structures at multiple levels (Olds, 1998 and Armstrong et al., 2011). In rare cases, pre-event host communities have strong legal support (Sadd, 2010). More commonly, they are heterogeneous and relatively disadvantaged – by virtue of economic resources, youth, minority or migrant status – when it comes to forming coalitions and lobbying to defend their position (Hall, 2005, Shin and Li, 2013 and Smith and Himmelfarb, 2007). Some marginalised groups have made effective use of media coverage to advocate for their interests, lobbying on compulsory purchase, housing, employment and pay (Fussey et al., 2011, Sant et al., 2013 and Vigor et al., 2004). Nevertheless, such resistance faces a massive challenge.

The interests of the urban growth machine are enforced through exceptional planning powers, authorising demolitions and the displacement of people (Andranovich et al., 2001, Boykoff, 2014, Gray and Porter, 2014, Hillar, 2003 and Lenskyj, 2000). Property developers and owners, including the relatively affluent middle classes, tend to be prioritised by local politicians and media over those of generally poorer, local communities (Gruneau, 2002). Although protecting vulnerable communities or ultimately offering improved social or affordable housing may be the stated aim, some commentators have analysed exceptional planning powers as a manifestation of neoliberal urbanism, with class- or ethnically-based 'cleansing' enforced as a prelude to the marketisation of urban space (Harvey, 1989, Kallin and Slater, 2014, Nam and Seok, 2008, Souliotis et al., 2014 and VanWynsberghe et al., 2013). Within this framework, the mega-event is an 'alibi' (Boykoff, 2014), facilitating a gentrification process: less desirable urban residents can be moved out; more affluent residents moved in; increases in land and property value are captured (Lauermann, 2015 and Watt, 2013). Policy and media narratives can be complicit in this process, systematically stigmatising already vulnerable people and places and framing the regeneration intervention as the solution to a problem which they have defined (Gray and Mooney, 2011 and Thompson et al., 2013).

Pre-event activity around venues and the Athletes' Village typically includes displacement of resident populations and housing demolition, to the extent that forced evictions are an expected part of the process (Olds 1998), and have been identified in relation to 32 different mega-events since 1980 (COHRE, 2007). Environmental improvement projects prior to the Beijing 2008 Olympics were described as 'a euphemism for demolition and displacement', with an estimated 1.5 million people displaced (Centre On Housing Rights and Evictions (COHRE), 2007 and Shin and Li, 2013: 559). In other instances, environmentally protected land has been reclassified, and smaller businesses are particularly vulnerable to clearance (Follman, 2015 and Raco and Tunney, 2010). The 'host community' at the time of the event may, therefore, be markedly different from the community in existence at the time of the bid.

## 2.2. Event time: experiencing securitisation

At the event time, a growing list of behaviours or processes have been defined as security concerns, requiring surveillance or control, with this 'securitisation' meaning that living around a mega-event is increasingly likely to involve constraints (Bennett and Haggerty 2011:4; Fussey et al., 2011 and Samatas, 2011). Presenting a carefully sanitised image of the city for television audiences, international visitors and corporate interests is an established phenomenon in mega-events, leading to concerns about infringement of civil rights and the impact of the security agenda upon local communities (Graham, 2012, Hall, 2005, Houlihan and Giulianotti, 2012 and Newham Monitoring Project (NMP), 2013).

Additional to pre-event evictions, 'cleaning operations', harassing or detention of minority groups including the homeless, migrants, street children and gay people, have been recorded in numerous host cities (COHRE, 2007).

A military presence within the host community may also be required (Fussey et al. 2011). Security for the London Olympics was boosted to include 18,200 military personnel (more than were deployed in Afghanistan at the time), additional to police and private security services, with the armed forces comprising up to half of security staff in and around the Games venues (Corera and Heald, 2012 and Houlihan and Giulianotti, 2012). The Newham Monitoring Project (NMP), a community based civil rights organisation in one of the London 2012 host boroughs, documented a 'climate of fear' associated with the intensive security around the Games, which included missile launchers, fighter jets and helicopters overhead, an 11-mile long electrified fence and additional electronic surveillance (Newham Monitoring Project, NMP, 2013: 2).

## 2.3. After the event: economic displacement

In the post-event period, the host community can find that the promised legacy is targeted towards more affluent incomers and non-residents. The main economic benefits from the event may be felt in other parts of the city. Where there are jobs for local people, these might be temporary or of poor quality, leading to frustration and disappointment within the host community (Boyle et al., 2008, Gray and Mooney, 2011 and Newman, 1999). Mega-events are said to function as a means of gentrification that 'permanently place housing beyond the financial means of a significant segment of society' (COHRE, 2007: 11). The mega-event village is an embodiment of the 'new' urban image, the housing generally being for new, more affluent people (Muñoz, 1997 and Sadd, 2010).

Furthermore, there can also be a 'ripple' effect in housing costs beyond the Village, with rising land and property prices resulting in increased rental costs, pricing out those on low incomes (Sadd, 2010 and Pillay and Bass, 2008). Host community members can be further marginalised as familiar neighbours leave, support services close, and older businesses or social hubs are replaced by amenities more appropriate for the new, more affluent community (Marcuse, 1985: 208; Varley, 1992).

'White elephant' facilities, appropriate for a major international festival but not sustainable on an ongoing basis, are one of the most notorious risk factors in relation to mega-events (Gold and Gold 2011). However, there is also the possibility that relatively well-used venues may serve only more affluent citizens and not be financially accessible to the host community (Clark and Kearns, 2015). Similarly, transport improvements often favour more affluent members of society, such as drivers or train users.

## 3. THE GLASGOW 2014 COMMONWEALTH GAMES

The challenge of evaluating legacy within any given context is exacerbated by the diverse nature of mega-events themselves. This being the case, any consideration of potential legacy, either positive or negative, should avoid simplistic inferences about events which may be qualitatively quite different. Although not one of the largest mega-events, the Glasgow 2014 Commonwealth Games (GCWG) is still of considerable scale and can be considered a critical case through which the potential of a mega-event to benefit the host community might be explored. The host city has well-developed governance and partnership structures with considerable experience of event management and 'boosterist' activity (Boyle et al., 2008 and Garcia, 2004). From the time of the successful bid in 2007, local and national government worked together with the explicit aim of leveraging the event as a regeneration initiative, in accord with good practice that there be an early focus on legacy, embedded within pre-existing urban strategy (Law, 2002, Matheson, 2010 and Vigor et al., 2004); in this case, the event occurred one third of the way into a 25 year regeneration project. In short, should things*not* go well for the host community in this instance, it could be said that there is little hope of positive outcomes in the case of other cities and circumstances.

## 3.1. Dalmarnock: a critical place at a critical time

The host community of Dalmarnock is a historic area of Glasgow 4 km east of the city centre. From the postwar period, Dalmarnock suffered from deindustrialisation, population decline, deteriorating housing stock, and increasing unemployment (Barke and Sim, 1981 and Rich, 1981). By 2012, after the demolition of four tower-blocks a decade earlier, the area included around 700 households, mostly social renters, living amidst considerable vacant and derelict land. Relative poverty and poor health have led to stigmatising media coverage of Dalmarnock as a 'problem place' inhabited, by extension, by problem people (Mooney, 2009: 442).

Dalmarnock is the site of the £230 m Athlete's Village, retrofitted to provide 400 homes for rent and 300 homes for private sale. Flagship venues for the GCWG, the Emirates Arena and Sir Chris Hoy Velodrome ('Arena' and 'Velodrome') were built to the north-west of the area (See Fig. 1 and Fig. 2). Land clearance practices in the pre-Games period have been criticised, since private land owners were compensated on the basis of potential land value following development while private householders received only current property value (Gray and Porter, 2014). The high-profile eviction of one owner-occupier resident (Mrs Jaconelli) placed much media attention on the area. The family concerned was isolated for five years during a dispute over compensation, and after neighbours in social rented accommodation were moved out of their housing block (Porter et al., 2009). The block previously included a Post Office and the last local shops in the area, which were lost to the community.

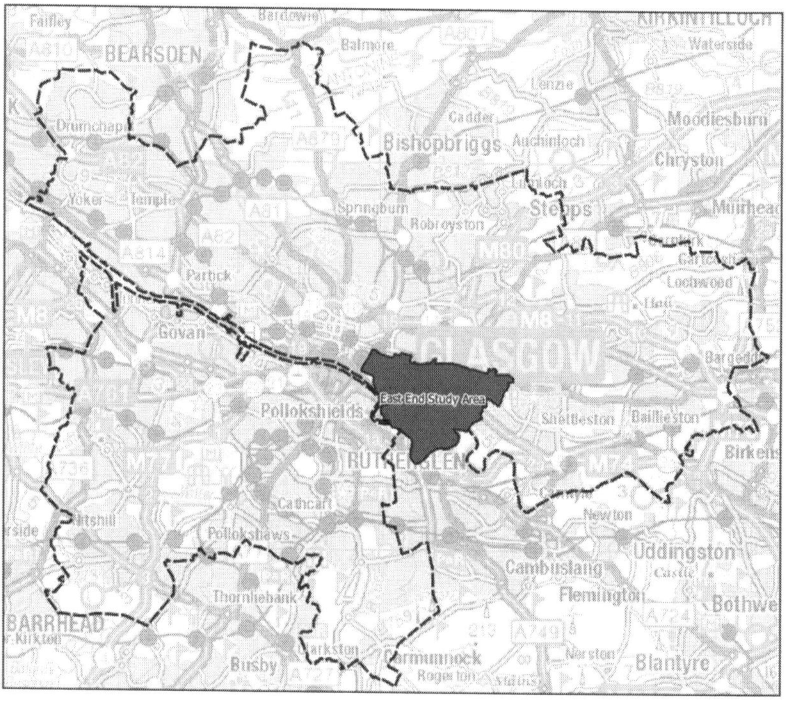


**Figure1.** Glasgow City and the wider East End study area.

## 4. RESEARCH AIMS AND METHODS

Focusing on Dalmarnock, where the greatest level of intervention in relation to the 2014 Commonwealth Games took place, we sought to explore host community experiences and perceptions in relation to:

- Reflections on the past
- Views around the present
- Hopes and fears for the future

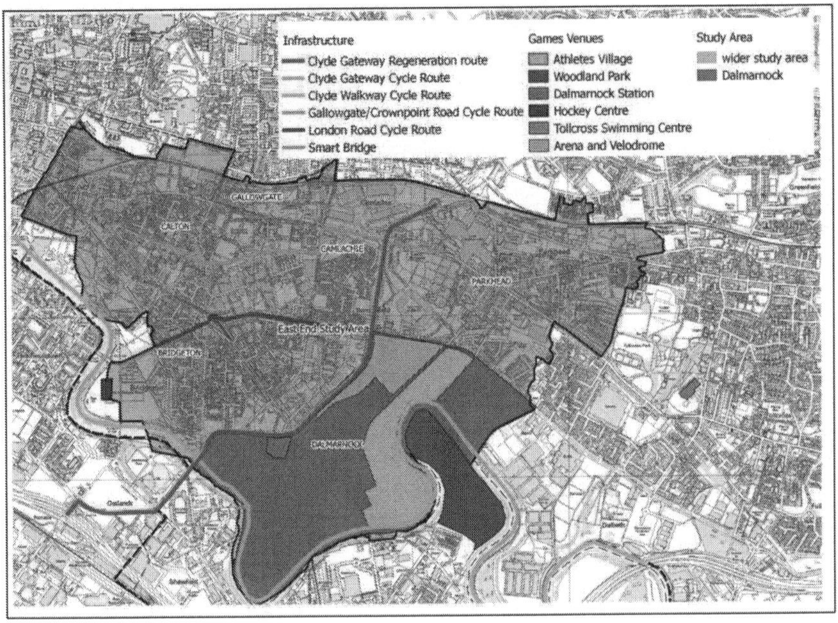


**Figure 2.** Dalmarnock and Commonwealth Games-related regeneration developments.

## 4.1. Data collection

Interviews with Dalmarnock residents were undertaken in May–July 2014, immediately preceding the GCWG. The timing was planned to capture experiences of living with a mega-event before the event itself coloured perceptions. Given the substantive relevance of gentrification processes, interviews were conducted with people responsible for paying the rent or mortgage, since they would have most insight into locational decisions. Initial contact was made by letter, approaching social housing residents in the area and others who had previously participated in social research. In order to take account of the changing community profile since the beginning of the regeneration, we used a snowball sampling method to engage residents with a variety of knowledge of, and perhaps, commitment to the area (Table 1).

**Table 1.** The research sample.

| Category | Description | Pseudonym |
|---|---|---|
| Long-term Residents: Resident in the area from before 2007 | Life-long: Lived only in Dalmarnock. (n = 2) | Barbara<br>Fiona<br>Angus & Iain (son)<br>Donna<br>Alan & Maria<br>Jon |
| | Established: Long-term Dalmarnock resident. Also, periods of residence elsewhere. (n = 10) | Linda<br>Eleanor<br>Alex<br>Deborah<br>Jim<br>Sandra<br>Susan |
| Recent Residents: Moved to Dalmarnock from 2007 onwards | Returning: Recent Dalmarnock resident. With some previous residence in Dalmarnock or the east end of Glasgow. (n = 4) | Mark<br>Anne & Dave<br>Rob & Cindy<br>Lucy |
| | New: Recent Dalmarnock resident. No previous Dalmarnock or east end residence. (n = 5) | Marion<br>Jeff<br>Michael<br>Catriona |

'Long-term' residents had lived in the area for at least the last 7 years, since the advent of Clyde Gateway and the award of the GCWG. This group comprised 'lifelong' residents, who had always lived locally and other 'established' residents, who lived in other areas before moving to Dalmarnock. 'Recent' residents were either 'new', with no previous residence in Dalmarnock or east Glasgow, or 'returning' interviewees who had moved to Dalmarnock within the past 7 years but had prior knowledge of the area, having lived previously either in Dalmarnock or elsewhere in east end of the city.

Semi-structured interviews were conducted, exploring resident attitudes, perceptions and experiences while retaining a level of comparability between accounts (Arthur and Nazroo, 2003). Twenty-five research participants from twenty one households in the area were asked to reflect on community change in the time that they had lived in the area, on changes associated with regeneration and the GCWG, and their expectations of the future.

## 5. REFLECTIONS ON THE PAST

Reflections on the nature and causes of neighbourhood change explored residents' perceptions of the role of the GCWG. As illustrated below, interviewee perspectives on change in Dalmarnock were informed by their level of familiarity with the area and by age, with a clear division between the *new* resident group and other participants. Following this, neighbourhood composition, displacement and loss of amenities are explored under the theme of *demolition and its consequences*.

## 5.1. New impressions

The *recent residents* were consistently positive about regeneration and the GCWG, focusing on the new or improved sports facilities in the area and the refurbished train station, which had recently reopened. For this group, regeneration and the GCWG were an attraction to the area. Lucy and her husband bought their home, regarding the area as 'up and coming', intending to let the property in the future if they were to move. Jeff and Michael were both social renters, pleased with the quality of their housing association properties as well as the change in the area. Marion owns a home in a mixed-tenure development and though change had not yet gone far enough. She was disappointed by the failure of a proposed hotel and retail development on vacant ground close to the Arena. Although resolute that she lived in a working class area and that the nature of the area was not starting to change ('Not at all'), Marion would prefer:

...improved transport facilities, doctors surgery and pharmacy, more of a choice of normal shops, getting rid of the pathetic pubs, a hotel or sort of bar that did, was more middle class. (Marion, *new*).

Michael (*new*) agreed that more should happen, with the proviso that local people 'buying in' was crucial. Despite the association of gentrification and 'pacification by cappuccino' (Zukin, 1995: 28), these sentiments seem quite far removed from revanchist urbanism (Smith, 1996). Nevertheless, Lucy thought about changes in people as well as the environment, saying there were fewer 'scummy people and neds[1]' hanging around before saying she felt safer in the area, partly connected to increasing familiarity and making friends in the neighbourhood. Optimism about the neighbourhood's future and a sense that physical changes have made the area look 'less depressing' were the dominant emotional impressions given by *recent residents*.

## 5.2. The long view: signifiers and drivers of neighbourhood change

Despite the intensity of recent demolition and construction activity, only three of the *long-term residents* focused on the GCWG as the main driver of neighbourhood change, with the rest having divergent perspectives on how, when and why their neighbourhood started to change.

Most *long-term* interviewees identified physical factors as the primary signifier of change in Dalmarnock, particularly the demolition of social housing some years before the GCWG bid. Eleanor and Alan (both *established*) described cycles of demolition and building, including the refurbishment and subsequent demolition of social rented homes in the early 2000s. Given these experiences, it is easy to see how local residents might be skeptical of potential legacy.

Fiona, (*lifelong*) suspected that earlier property demolition in the area may have been connected to later developments. Added to this was the perception that changes were 'not for the benefit of the community' (Jeff, *new*) and that 'a lot of people are getting their pockets lined' (Barbara, *lifelong*).

However, de-industrialisation from the 1960s onwards was considered the driving force of neighbourhood change by some older residents: as industry closed down, people followed employment opportunities and, as part of a slum clearance programme, much of the population was moved to peripheral housing estates. For these participants, the GCWG is therefore understood within a historical context of heavy industry and poor housing in an area that was 'Black, grimy, foggy. Always dark and foggy. A lot of smells...' (Cindy, *returning*).

## 5.3. Demolition and its consequences

Demolition and its consequences formed the other major theme, as participants reflected on neighbourhood change.

### 5.3.1. Neighbourhood composition

Many interviewees identified the demolition of high flats in the period before the GCWG bid, as a pivotal time for the area. Such large-scale clearance is generally believed to be damaging to communities (Paris and Blackaby, 1979). However, participants here were supportive about the demolition of the flats and the subsequent relocation of residents, believing it had changed the composition of the neighbourhood for the better. Alex associated the demolitions with what might be considered 'cleansing' practices common in the run up to a mega-event (Grix, 2013), moving out 'hooligans...all the troublemakers' (Alex, *established*). However, for one couple, this displacement, alongside regeneration activity, attracted them back to neighbourhood after living elsewhere:

That's why we decided to come doon here because we heard they were making it a better area an' they threw a' the riff raff oot. (Dave, *returning*)

The dismissive language above could be considered as 'distancing' language, used by the speaker to differentiate themselves from stigmatised populations or places (Wacquant, 2007a). However, other interviewees were more delicate in communicating the same idea while speaking of the neighbourhood being quieter and safer.

It seems, at least from the perspective of those that remain, that disorder rather than poverty or class may have been at issue and there was consensus over the neighbourhood now being a better place to live than 15–20 years ago. Furthermore, Dalmarnock has attracted both the *new* interviewees and *returning* participants who either have knowledge of the area through previously living there, or are returning residents after time elsewhere, seems to reinforce the point.

### 5.3.2. Displacement

With regard to more recent GCWG-related social change, two interviewees mentioned the Jaconelli eviction. Despite acknowledging that 'fair enough, it was her family home', on balance, Michael (*new*) took the view that 'those flats down there had had their day. It was time to go'. For Jon (*established*), the heavy-handed nature of the eviction, with 'over a hundred police to get rid of her and her husband from one flat', provoked a distrust of the authorities involved:

I mean, you know, if they do all that stuff under the glare of publicity, you know? (Jon, *established*)

Jim and Susan both belong to Show People communities,[2] who have longstanding ties with Dalmarnock. Many were displaced by recent regeneration and now live in other areas, some abandoning their traditional lifestyle. Susan (*returning*) had to relocate but managed to stay within the area. She was unhappy with the reason given for the relocation, that caravan sites would be an 'eyesore' for tourists. In the case of Jim's site, pressure to relocate seems to have been successfully resisted:

We, like anybody else, kicked up a lot of stink, and one thing and another, finish up, not one of us has moved. (Jim, *established*)

While Mrs. Jaconelli's eviction did not feature strongly in resident accounts and ultimately some of the Show People have secured a long-term future in the area, these aspects of displacement reinforce concerns around mega-events and the treatment of vulnerable and minority groups (Bernstock, 2014, Gray and Porter, 2014 and Watt, 2013).

### 5.3.3. Loss of amenities

The closure of small, local shops and a post office, in order to make way for the new Athletes' Village ('the Village'), was a widespread concern in the interviews. Although there is a very large supermarket to the south-east of the area, this was considered, by some, to be an inaccessible and expensive option. A consequent need for car transport was emphasised and some interviewees described elderly or less physically able neighbours resorting to taxis, a costly option for people on low incomes. Although, intermittently, there was temporary shopping provided initially, by Clyde Gateway and, later, a local entrepreneur, the value of local shopping to the social life of the area, as well as for practical purposes like the collection of pensions or small, daily shopping needs, came across clearly in the interviews. The loss of local shops, children's play areas and the closure of the Dalmarnock community centre also had negative impacts on socialisation, day-to-day exercise, and building social support in the area.[3]

## 6. VIEWS AROUND THE PRESENT

There is a tension between the long-term process of regeneration and the fixed deadline of a mega-event. To ask about attitudes towards the event and associated developments in the host community is, necessarily, to ask about the whole cluster of experiences around living with that event. Views on the present, therefore, are divided into reflections on event-related regeneration processes, and experiences of or opinions about the new developments at the time just before the event.

## 6.1. Processes of regeneration and event management

### 6.1.1. Eleven days of sport, five years of dirt and disruption

The physicality of the regeneration experience is a recurrent theme, recounted directly by established community members and second-hand by more recent residents who had been less affected in the preceding years, but sympathised with their neighbours. In some cases, interviewees were relatively tolerant about disruption, mostly because it was considered a necessary evil:

Changes now, I think, are absolutely horrendous but understandable. (Barbara, *lifelong*)

They have been... well no' too much disruptive. It's just there's been a lotta inconvenience. (Dave, *returning*)

However, two of the established interviewees, Donna and Linda, revised their views on the GCWG from initially supportive to hostile, due to disillusion with the closure of shops and the community centre, increasing disruption and a sense of being disregarded by the authorities.

Demolition, digging up the foundations of old buildings, construction of roads, houses and facilities resulted in noise, vibration and continual dust over a protracted period, causing damage to cars and property, requiring repeated cleaning. Construction lorries were another source of noise and a cause of safety concerns. Alan was concerned about health risks because of contamination from old industrial works being disturbed, saying 'there's an awful lot of people getting cancer an' that livin' down here' (*established*). For both visitors to the neighbourhood and residents, the amount of physical change in the area was described as 'disorientating'.

### 6.1.2. Restrictions and rumours

The disorientating effect was amplified in the weeks before the GCWG, with the erection of security fencing and barriers restricting movement along different roads and pathways, which attracted the most widespread and heartfelt criticisms. Beyond the practicalities of negotiating re-routed transport and around blocked streets or pathways, there were issues of status and respect, including: being on 'lockdown', a term otherwise applied to prison governance; being unable to park near your own home; knowing that you are not one of the 'important' people who have access to dedicated transport lanes; or being monitored by security guards and cameras and feeling imprisoned:

As I say, the best thing they could've done is just put a big cage right roond the whole lot, gave you a horse's bag and told you to go oot there an' feed yersel'. That's how good it's got. (Angus, *established*)

Several participants felt that, recognition of the challenges of living with a mega-event was completely absent beyond the community; Deborah exclaimed 'if they'd even have paid us any kind of lip service, even' (established). Interviewees who had been told that they might get a one-off cleaning service or car wash token reacted with derision about the usefulness of the offer. The idea that they were living though a short-term inconvenience caused considerable resentment (Alex, established).

Rumours were a recurrent theme in relation to GCWG security restrictions and there was considerable criticism of the level and nature of communication with residents along with a view that no-one was responsible for coordinating between the GCWG and the community:

Nobody fae the Council are coming and telling us exactly… all they'll tell is 'see when it's aw finished, you'll no' know the place' (Laughs). (Sandra, *established*)

A lack of clarity over exactly what would happen and what the implications of the restrictions would be for local residents led to anxiety and discontent: some people had heard that there would be a curfew in place, that children could not go out in the street after 6 pm or that playing music after 9 pm would be prohibited in case it disturbed the athletes or that the petrol station would not be able to sell petrol while the GCWG were on. Interviewees experienced uncertainty about being able to get to work and access to health services or the shops, either for themselves or on behalf of their neighbours. Most seriously, in an area including many people with health problems, interviewees spoke of security barriers disrupting ambulance and healthcare worker access.

Some residents mentioned or had attended local meetings put on by the authorities, but there was also anger and frustration that the purpose of these was perceived to be delivering decisions that already been made, rather than listening and making decisions based on local community input. People's experience led to scepticism over the place of the community in 'the pecking order' (Deborah *established*) and may have contributed to mixed views about the Games and a polarised atmosphere at meetings. Michael (*new*) described being at a community meeting where 'at least two people have said to my face 'this is a fantastic thing' and at the community meeting shouting 'bloody Commonwealth Games'. At some level the GCWG may have succeeded in building community capacity in ways not quite anticipated; Linda contributed what might be considered the most positive perspective on meetings associated with the GCWG:

I think wi' all this it's brought people closer the gither, 'cause everybody's in the same boat... everybody's got the same anger. You know, at what's happening tae them. (Linda, *established*)

## 6.2. New infrastructure and amenities

Interviewee experiences and attitudes towards the new infrastructure, amenities and housing were predominantly supportive, although participants who were least supportive of the Games offered surprisingly little commentary on regeneration developments, other than the housing, which tended to be positively regarded.

### 6.2.1. Infrastructure developments

With regard to infrastructure improvements, the recently reopened train station was praised as providing a good service, being well-used and also feeling safer than before. None of the participants mentioned cycling personally, but Fiona (*lifelong*) was pleased that the new cycle-way provided a safe route for her son to cycle to the new Arena. Others commented on watching people cycling in the area, including across the new bridge across the river Clyde connecting Dalmarnock to South Lanarkshire. Several participants mentioned the new roads favourably. For Lucy (*new*), they were one of the attractions in moving to the area. Catriona (*new*) does not drive but reported that visitors and neighbours had found them a great improvement. However, there was widespread dissatisfaction over 'shocking' bus services, which are a more important mode in less affluent communities.

## 6.2.2. Sport and leisure facilities

Some commentators have questioned the value of the new sport and leisure facilities to the local community, particularly, whether a velodrome is a useful amenity (McCartney et al., 2012). Jeff (*new*) shared the view that the facilities are not really for local people, ridiculing the cost of the Arena spa. Alan (*established*) also thought that the gym and velodrome might be quite expensive but said that a lot of local children used them.

However, aside from the spa component, participants were generally pleased with other facilities, which were considered attractive and well built. *New* and *returning* residents were most likely to enthuse about the developments. Of the newer residents, Michael, Rob and Cindy were particularly enthusiastic, with the latter two having visited all of the new amenities close to the area. Michael described the arena and velodrome as 'fantastic structures', as well as mentioning that his daughter and her friends liked to go there to watch the boys playing five-a-side football.

Overall, *lifelong* and most of the *established* residents also considered the facilities positively. Fiona (*lifelong*) said her son went to football classes at the Arena. Jon (*established*), although noting that the new facilities were of less use to older residents, planned to cycle at the Velodrome, while Mark and Deborah also knew family and people in the area who had already used the gym and velodrome. Alex and Deborah, who had been especially critical of some aspects of the

GCWG and regeneration, described the Arena and Velodrome as 'amazing' and 'a fantastic asset to the area'. Other interviewees spoke about the new woodland park development across the river as a welcome addition for local children. The redevelopment of the previously derelict 'Olympia' building in neighbouring Bridgeton also attracted approval, as did the re-opened library and new mediatheque inside.

### 6.2.3. The Village

Of all the new developments in the area, the Village attracted most enthusiastic discussion. Considering the GCWG overall, Angus and Linda were amongst the most critical interviewees. Angus was against the city hosting the Games from the beginning, believing money should not have been spent on shops and homes rather than new, prestige sports facilities:

I'd say the Commonwealth Games should've been in Edinburgh because Edinburgh's got it a' whereas we've not. (Angus, *established*).

However, he was amongst several interviewees who said that they or their relatives would like to live in The Village if a home became available. Similarly, Linda was pleased to get a leaflet inviting her, along with other neighbours, to go to a viewing of the Village properties. As well as being impressed with the quality of the houses, she was pleased about new homes being built in the area. The aesthetic appeal of the houses was praised as 'lovely' and 'amazing' as well as 'high-spec', and the reuse of what was previously wasteland was a cause of satisfaction.

# 7. HOPES AND FEARS FOR THE FUTURE

Hopes and fears for the future clustered around two main themes: overcoming stigma and the sustainability of positive changes in the area; and the extent to which the Village and new amenities would benefit the local community.

## 7.1. Stigma and the sustainability of improvements

A sense of the neighbourhood changing for the better was often grounded in reflections of a more challenging past. Older residents, in particular, spoke about historical deprivation and the impacts of stigma. A 'blemish of place' (Wacquant, 2007b: 67), based on the area's reputation for violence, gang fighting and muggings, had become associated with the people who lived there, making it difficult for them to find work:

You just said you came fae somewhere else. Rutherglen, Lower Rutherglen. That was a good one you used to tell...Anywhere but Dalmarnock! (Alan, *established*)

Several residents shared experiences of previous, casual stigmatisation, as passing strangers and even friends made judgements based on the appearance of the neighbourhood; Linda was not alone in finding it hurtful and unjust that a poorly maintained neighbourhood would reflect badly on anyone who lived there:

You know, it's like, 'Oh, look at the state of they fences, imagine living in that, it's like Chicago' and all this, you know? And you don't... no, it's no' like that. It is nae like that, you know?...You hear them on the bus coming on, "I would nae like tae live in there." And yet the people are absolutely fantastic...You know, and I think people make a place. (Linda, *established*)

Area stigmatisation was not entirely relegated to the past. Supporting Gray and Mooney's research (2011), participants criticised the media for projecting a misleading image of the area, conflating low income with moral degradation and ignoring successes. Rob spoke about his daughter, who was brought up in Dalmarnock when he lived there years ago:

She's a nurse. But you don't hear the good things; they were wanting certain people to come down to the Olympia to gie something negative about living in Dalmarnock. I said there, 'My daughter got a degree. Do you want to hear about that?' (Rob, *returning*)

However, some interviewee accounts included evidence of stigma reversal prior to the GCWG. Michael spoke about being proud to put 'Dalmarnock', rather than just 'Glasgow' when writing his address. Similarly, Fiona, one of the *lifelong* residents, said that now she tells people she is from Dalmarnock, when previously she would name a neighbouring area because she had felt Dalmarnock was a bad place to come from, with a reputation as a 'rough' area. The more enthusiastic advocates of recent change in Dalmarnock made comparisons with the west end of the city and more affluent suburbs, regarding the regeneration as a means of reducing environmental inequalities.

Interviewees valued higher levels of neighbourhood maintenance and a perception that the area's reputation had improved as a result. Some associated these benefits with the security and surveillance aspects of the GCWG, which were not viewed wholly negatively. New security cameras were associated with lower levels of littering and feeling safer, while the attention paid to street cleansing in advance of the GCWG in the neighbourhood was described as 'fantastic':

The only way I can describe it before when I first moved here is dirty...I find it very clean now. To be quite honest with you, it's about 17 days to the Games. The council are nothing short of hoovering that street every day. (Michael, *new*)

However, there were also fears that investment in and the gains to neighbourhood reputation and pride might be lost following the Games; both the place and the people who live there might again be forgotten. In particular, there was also a perception that private interests would continue to be a priority over the needs and wants of local people. Jim observed of the new river walkway:

After the Games, that walkway, if there's a developer comes and he wants so many square feet, and that walkway's in the road, that walkway will disappear. (Jim, *established*)

Bearing in mind previous cycles of demolition and construction it is unsurprising that, although people welcomed care and attention to their neighbourhood and improvements to the reputation of the area, there was a far more cautious attitude when considering whether positive changes would be sustained over the longer term.

## 7.2. Community or communities?

Looking to the future, new housing and the return of shops to the area were the dominant themes. There was broad agreement that new homes were needed in the area and several participants also saw the Village, transport infrastructure and the new venues as potential mechanisms for revitalising Dalmarnock:

... it looks lovely and, again, it's just waiting (the new Village area) for those to become full of people and hopefully it will sort of turn around, that Dalmarnock will become more populated and hopefully we'll get more shops. (Barbara, *lifelong*)

There was some agreement that new roads would encourage business and industry to locate in the area. The new venues were also seen as attractions to the neighbourhood, both for community use and in bringing employment opportunities, offering benefits to social and psychological well-being.

The issue of tenure in the Village was important to participants, with opinion favouring social rented or mixed tenure developments. Alan took the view that, without the GCWG, all of the land would have been sold to developers and used for private housing which:

... would nae have done anything for the community. Know what I mean? A lot of the houses that are bought, people in them go to their work, come hame and that's in. (Alan, *established*)

However, Jon (*established*) felt that having a private component in the housing mix enabled local people who want to buy a house to stay in the area. Like some of the older participants, he reflected on what the loss of more affluent community members meant for sustaining services in the area when tenure is 'not representative':

It affected me, you know, people on a low income that weren't, you know, didn't have loads of money to spend in the area and so... So, certainly, you know, I was strongly in favour of, you know, new build housing coming in that would have enabled... people to stay. People that, you know, had money to spend in shops in the area, to stay and live and, you know, be with their families and all the rest and know in the neighbourhood... (Jon, *established*)

Rather than understanding the planned mixed community in terms of gentrification (Gray and Mooney, 2011), interviewees with a longer perspective considered a mixed community in Dalmarnock, including high quality housing, as a return to the past:

...this is going to be a better class of people, the way it was then. Because in the red buildings that was starter people, you know? They were training to be teachers, owner-occupiers, so that was like their starter houses, so whether they were training to be engineer or whatever. Once they were finished their apprenticeship they would move out the area but that was their start... (Cindy, *returning*)

However, there were two qualifying concerns to general support the mixed tenure Village, relating to affordability and the implications of the Village for community cohesion in wider Dalmarnock.

Affordability was raised as an issue in relation to both renting and buying locally. Rob and Cindy (*returning*) were originally on the list to move to rented accommodation in the Village but, although still enthusiastic about what the Village might do for Dalmarnock, they accepted a property elsewhere in Dalmarnock, considering it still high quality but less expensive to rent. Mark (*established*) argued that new properties should *only* be available for social rent, on the grounds that 'if you can afford tae buy a house, you can afford tae buy it anywhere', and although, the houses for sale might represent very good value, for families on relatively low incomes they are still unattainable. New residents, Lucy and Catriona both regarded their homes as investments, hoping that their value will 'improve'. Although Lucy mentioned possibly renting their property out in the future, both she and Catriona also seemed connected to Dalmarnock at a deeper level, engaging with neighbours and spending time in the area.

With respect to future community cohesion, there were concerns about who future tenants might be, in both the private and social rented sectors. Cindy described one housing association as:

... only letting to working couples...or (people with a pension). Well, people that are paying their way, you know? Yeah. Some of them are saying that's discrimination for those people on housing benefit... (Cindy, *returning*)

However, a lack of any vetting procedures was also a concern, with regard to private tenants in buy-to-let investment purchases and social renters. There were fears that without any controls, residents in the current community could end up living next to 'a thug' (Alan, *established*).

The wide variation in housing quality within a small area was seen as a potential strain to relations between people in the Village and those in wider Dalmarnock. Some of the housing stock in the area is still very poor, including properties adjacent to the new Village. Angus talked about cold and dampness, saying 'the mortuary's far warmer than this house', and described years of waiting for a new home. There was a sense of inequity that long-term Dalmarnock residents got 'the very cheapest stuff that they could get' in housing refurbishments while:

...I've had a wee look in the Athletes' Village, and they're fabulous... they're lovely, high spec, you know, really, really, quality stuff and... you know, so that's a wee bit frustrating. (Deborah, *established*)

Nevertheless, for many interviewees, the defining measure of success with the regeneration will not be the Village, the venues, new employment or the Games themselves, but the return of facilities essential to social interaction in the community, particularly small, local shops.

...a whole new community in and nae facilities. (Linda, *established*)

# 8. CONCLUSION

Latouche (2007) has described the phenomenon of the 'imaginary' legacy: mythologies of good and bad Games, created by sectional interests advocating for their own version of the 'real' legacy. Our investigation of the impacts and legacy of the Commonwealth Games 2014 for Dalmarnock, Glasgow shows that perspectives on legacy for the host community are a matter of scale, time and process.

With regard to scale, the summer Olympic Games can be considered the ultimate mega-event. However, smaller-scale '*not-so-mega-events*' are more commonplace and offer a more feasible opportunity for event-led regeneration for many cities; however, relatively little is known about their impacts ( Coates, 2012). A focus on 'mega' makes it easy to overlook the 'micro' – those local communities who are most impacted by the event. In the case of the Glasgow 2014 Games, local legacy related to the mega-event was defined in terms of improving the physical and social environment of the East End of Glasgow ( Glasgow City Council (GCC), 2007 and Scottish Government, 2014). Our in-depth research was conducted in what might be considered the 'core' host community, Dalmarnock, which was most affected by the Games and associated developments. Even though we have focused on one of the smallest communities in the East End of Glasgow, nonetheless, we find a mixture of residents and a variety of opinions. The complexity and heterogeneity of community views in respect of the mega-event is striking. Both between and within individuals, we did not find a singular outlook on the CWG and regeneration; participant perspectives on the Games and regeneration were nuanced, rather than falling into a binary 'for or against' pattern. The process of change attracted frustration and, in some cases, vehement negative commentary, particularly relating to shorter-term constraints at the time of the Games. Simultaneously, new infrastructure, amenities and especially housing were broadly welcomed, including by some participants who were very distressed by the regeneration process.

Time is relevant to studying legacy in a number of respects. A mega-event can be analysed in three distinct phases: the pre-event period when the bid and major preparations take place, such as the construction of new venues, and when, for more recent events, associated legacy programmes are put in place; the intensive event period, in weeks before, during and after the event, when arrangements for security, catering, accommodation and travel are put in place; and the post-event period when new facilities are converted for everyday use, with the continuation of some legacy programmes and monitoring. We encouraged local residents to reflect and found that individuals' views on the Games varied between each of these time phases. Nostalgia and disdain for the past could co-exist in complex patterns with discontent or mixed feelings about the pre-event phase, and, again, with both doubt and optimism for the future. People's horizons of memory coloured their boundaries of expectation: how people understood recent social change, and what they felt about the future, often related to the extent of their own history in the area. While some people attributed social changes to the advent of the Games and its associated developments, those with a longer cognitive time horizon viewed recent events as part of a longer process of social change. Rather than a gentrifying process, where lower income groups are displaced by the more affluent (Smith, 2012 and Thornley, 2012), these participants see the future in terms of a return to more mixed community, following a period where disruptive tenants were spatially concentrated through housing mechanisms.

As well as being an event realised in the moment, for the local host community, a mega-event is very much experienced as a process in several regards. Many of the interviews demonstrate emotional as well as physical impacts from living in the highly securitised environment just before the event and that these processes of securitisation are found alienating and disrespectful. While current mega-event literature tends to displacement and social change (Centre On Housing Rights and Evictions (COHRE), 2007 and Smith and Derkson, 2002), our interviews suggest that a more pressing issue for residents was inadequate communication, with authorities perceived to dictate, rather than consult on their practices. Another is insufficient recognition or acknowledgement of how onerous long-term co-existence with mega-event regeneration can be, given its intensity and scale. Furthermore, that, for those living with a mega-event, it is not adequate to answer these difficulties by directing residents to look to the future.

This analysis also highlights concerns about market-based processes of change, notwithstanding that the overall programme was initiated by public policy. Although we have found elsewhere that residents in the East End of Glasgow exhibit relatively high levels of perceived community empowerment (Clark and Kearns, 2013), in-depth analysis shows that residents in Dalmarnock feel powerless and concerned about two particular aspects of local change. Firstly, they were unable to prevent the loss of their local shops or secure their speedy return; the latter is dependent upon regeneration providing a suitable market opportunity for a general retailer. Secondly, people are well aware of the rapid growth of the private rented sector in the city (Scanlon et al., 2013), which they see as inadequately regulated, and were concerned to avoid concentrations of disruptive tenants, either in new social housing in the area, or in the private sector, when new owner-occupied housing might transfer to the rented sector.

Earlier mega-event research has provided an important critique of the dark side of legacy, focusing in particular on disadvantage to the already disadvantaged (Porter et al., 2009). However, in common with numerous changing post-industrial areas, Dalmarnock includes long-term residents, living alongside newer, incoming residents, and returning residents. In seeking to avoid assumptions about the different residents who have a stake in the area (Ahlfeldt et al., 2012 and Latouche, 2011) we adopted an inclusive approach to sampling. Although the use of mega-events for regeneration is likely to remain contentious, from the broad host community perspective, our findings offer some cause for optimism. As Somerville (2011: 16) observes, 'strong attachment to place does not have to be expressed through claims of exclusive ownership', and in fact we found considerable common ground in residents' hopes for the future in terms of a new, mixed community in this case. Nevertheless, interviewees' optimism was qualified by concerns about whether future benefits would be sustained and accessible to all; these were matters on which residents wanted a commitment from those responsible for change.

## ACKNOWLEDGEMENTS

We would firstly like to take this opportunity to thank all of the research participants, who kindly shared their experience and opinions. Secondly, we would also like to thank Breannon Babel, Camilla Baba, Joanne Neary and Steve Rolfe, who were a vital part of the research team, making connections and conducting interviews, along with Laura Baggley for her invaluable administrative support. *GoWell: Studying Change in Glasgow's East End* project is funded by the Scottish Government, NHS Health Scotlandand **SPORT**scotland. Claire Cleland's input was supported by the Medical Research Council as part of the Neighbourhoods and Health Programme (MC_UU_12017/10).

# REFERENCES


1. Ahlfeldt, G. M., Wolfgang, M., & Ölschläger, M. (2012). Support for and resistance against large stadia: The role of lifestyle and other socio-economic factors. In W. Maennig, A. Zimbalist, & R. A. Woods (Eds.), *International handbook on the economics of mega sporting events.* (pp. 279–294). Cheltenham: Edward Elgar.

2. Andranovich, G., Burnak, M. J., & Heying, C. H. (2001). Olympic cities: Lessons learned from mega-event politics. *Journal of Urban Affairs, 23*(2), 113–131.

3. Armstrong, G., Hobbs, D., & Lindsay, I. (2011). Calling the shots: The pre-2012 LondonOlympic contest. *Urban Studies, 48*(15), 3169–3184.

4. Arthur, S., & Nazroo, J. (2003). Designing fieldwork strategies and materials. In J. Ritchie, & J. Lewis (Eds.), *Qualitative Research Practice* (pp. 109–137). London: Sage.

5. Barke, M., & Sim, D. (1981). Getting into gear: Urban renewal in the east end of Glasgow. *Geography, 66*(4), 315–319.

6. Bennett, C. J., & Haggerty, K. (2011). 'Introduction: Security games: Surveillance and control and mega-events'. In C. J. Bennett, & K. Haggerty (Eds.), *Security games: Surveillance and control at mega-events oxon* (pp. 1–19). Routledge.

7. Bernstock, P. (2014). *Olympic housing: A critical review of London 2012's legacy.* Surrey: Ashgate.

8. Boykoff, J. (2014). *Celebration, capitalism and the olympic games.* New York: Routledge.

9. Boyle, M., McWilliams, C., & Rice, G. (2008). The spatialities of actually existing neoliberalism in Glasgow, 1977 to present. *Geografiska Annaler: Series B Human Geography, 90*(4), 313–325.

10. Centre On Housing Rights and Evictions (COHRE) (2007). *Mega-events, Olympic games and housing rights: Opportunities for the Olympic movement and others.* Geneva: COHRE.

11. Clark, J., & Kearns, A. (2013). *GoWell in Glasgow's East End. Baseline Community Survey 2012.* Glasgow: GoWell/ Glasgow Centre for Population Health (54(s2), 175-187).

12. Clark, J., & Kearns, A. (2015). Pathways to a physical activity legacy: Assessing the regen- eration potential of multi-sport events using a prospective approach. *Local Economy, 30*(8), 888–909. http://dx.doi.org/10.1177/0269094215601993.

13. Coaffee, J. (2012). Policy transfer, regeneration legacy and the summer Olympic Games: Lessons for London 2012 and beyond. *International Journal of Sport Policy and Politics, 5*(2), 295–311.

14. Coates, D. (2012). "Not-so-mega events". In W. Maennig,& A. Zimbalist (Eds.), *International handbook on the economics of mega sporting events* (pp. 401–433). Cheltenham: Edward Elgar.

15. Corera, G., & Heald, C. (2012). London 2012: How was it for the troops? BBC News (12/08/ 2012). http:// www.bbc.co.uk/ news/uk-19228232

16. Essex, S., & Chalkley, B. (1998). Olympic Games: Catalyst of urban change. *Leisure Studies, 17*(3), 187–206.

17. Follman, A. (2015). Urban mega-projects for a 'world-class' riverfront: The interplay of informality, flexibility and exceptionality along the Yamuna in Delhi, India. *Habitat International, 45*, 213–222.

18. Fussey, P., Coaffee, J., Armstrong, G., & Hobbs, D. (2011). *Securing and sustaining the Olympic City: Reconfiguring London for 2012 and beyond.* Aldershot: Ashgate.

19. Gaffney, C. (2010). Mega-events and socio-spatial dynamics in Rio de Janeiro, 1919–2016.

20. *Journal of Latin American Geography*, *9*(1), 7–29.

21. Garcia, B. (2004). Urban regeneration, arts programming and major events. *International Journal of Cultural Policy*, *10*(1), 103–119.

22. Glasgow City Council (GCC) (2007). *People, place, passion: Glasgow 2014 Commonwealth Games Candidate City File Glasgow: Glasgow 2014.*

23. Gold, J. R., & Gold, M. M. (2011). From a to b: The Summer Olympics 1896–2008. In Gold,& Gold (Eds.), *Olympic cities 1896–2016 London* (pp. 17–54) (2nd ed.). Routledge.

24. Graham, S. (2012). 'Olympics 2012 security: Welcome to lockdown London'. *The Guardian* (12/3/2012 http:// www.theguardian.com/sport/2012/mar/12/london-olympics-security-lockdown-london).

25. Gray, N., & Mooney, G. (2011). Glasgow's new urban frontier: 'Civilising' the population of 'Glasgow East'. *City: Analysis of urban trends, culture, theory, policy, action*, *15*(1), 4–24. http://dx.doi.org/10.1080/13604813.2010.511857.

26. Gray, N., & Porter, L. (2014). By any means necessary: Urban regeneration and the 'state of exception' in Glasgow's Commonwealth Games 2014. *Antipode*, *47*(2), 380–400.

27. Grix, J. (2013). Quo Vadis SMEs? In J. Grix (Ed.), *Leveraging legacies from sports mega-events: Concepts and cases.* (pp. 178–179). London:: Palgrave Pivot.

28. Gruneau, R. (2002). Foreword. In M. Lowes (Ed.), *Indy dreams and urban nightmares.*

29. Toronto: Toronto University Press (ix–xii).

30. Hall, C. M. (1992). *Hallmark tourist events: Impacts, management, and planning.* London:Belhaven Press.

31. Hall, C. M. (2005). Selling places: Hallmark events and the reimaging of Sydney and Toronto.

32. In Nauright, & Schimmel (Eds.), *The political economy of sport* (pp. 129–151).

33. Harvey, D. (1989). From managerialism to entrepreneurialism: The transformation in urban governance in late capitalism. *Geografiska Annaler Series B, 71,* 3–17.

34. Hayes, G., & Horne, J. (2011). Sustainable development shock and awe? London 2012 and civil society. *Sociology, 45*(5), 749–764.

35. Hillar, H. H. (2003). Towards a science of Olympic outcomes: The urban legacy. In M. Moragas, C. Kennett, & N. Puig (Eds.), *The legacy of the Olympic Games 1984–2000* (pp. 102–209). Barcelona/Lausanne: International Olympic Committee.

36. Horne, J., & Whannel, G. (2012). *Understanding the Olympics.* London: Routledge.

37. Houlihan, B., & Giulianotti, R. (2012). Politics and the London 2012 Olympics: the (in) security Games. *International Affairs, 88*(4), 701–719.

38. Kallin, H., & Slater, T. (2014). Activating territorial stigma: Gentrifying marginality in

39. Edinburgh's 'other' fringe. *Environment & Planning A, 46*(6), 1351–1368.

40. Kasimati, E. (2003). "Economic aspects and the Summer Olympics: A review of related research." *International Journal of Tourism Research*, *5*(6), 433–444.

41. Lauermann, J. (2015). Temporary project, durable outcomes: Urban development through

42. failed Olympic bids? *Urban Studies*. http://dx.doi.org/ 10.1177/ 0042098015585460.

43. Latouche, D. (2011). Montreal 1976. In Gold, & Gold (Eds.), *Olympic cities 1896–2016.*(pp. 47–267) (2nd ed.).

44. Law, C. M. (2002). *Urban tourism: The visitor economy and the growth of large cities.*London: Continuum.

45. Lenskyj, H. J. (2000). *Inside the Olympic industry: Power politics and activism.* Albany: State University of New York Press.

46. Lenskyj, H. J. (2002). The best Olympics ever? Social impacts of Sydney 2000. *SUNY series on sport, culture and social relations.* Albany: State University of New York Press.

47. Leopkey, B., & Parent, M. M. (2011). Olympic Games legacy: From general benefits to sustainable long-term legacy. *International Journal of the history of Sport, 29*(6), 924–943.

48. Liao, H., & Pitts, A. (2006). A brief historical review of Olympic urbanization. *The International Journal of the History of Sport, 23*(7), 1232–1252.

49. Marcuse, P. (1985). Gentrification, abandonment, and displacement: Connections, causes, and policy responses in New York City. *Journal of Urban and Contemporary Law, 28*, 195–240.

50. Marcuse, P. (2015). Depoliticizing urban discourse: How 'we' write. Cities, 44, 152–156.

51. Matheson, C. M. (2010). Legacy planning, regeneration and events: The Glasgow 2014 Commonwealth Games. *Local Economy, 25*(1), 10–23.

52. McCartney, G., Hanlon, P., & Bond, L. (2012). How will the 2014 Commonwealth Games impact on Glasgow's health, and how will we know? *Evaluation, 19*, 24–39.

53. Mooney, G. (2009). "Problem' populations,'problem' places'. In J. Newman, & N. Yeates (Eds.), *Social justice: Welfare, crime and society* (pp. 97–128). Maidenhead: Open University Press.

54. Müller, M. (2015). What makes an event a mega-event? Definitions and sizes. *Leisure Studies.*

55. Muñoz, F. (1997). Historic evolution and urban planning typology of Olympic Villages Barcelona. *Centre d'Estudis Olímpics.*

56. Muñoz, F. (2006). Olympic urbanism and Olympic villages: Planning strategies in Olympic host cities, London 1908 to London 2012. *The Sociological Review, Special Issue: Sociological Review Monograph Series: Sports Mega-Events: Social Scientific Analysis of a Global Phenomenon, edited by J Horne and W. Manzenreiter, 54*(s2), 175–187.

57. Nam, S. W., & Seok, K. O. (2008). Olympics and housing: Urban renewal, forced evictions and marginalization of minorities. *Korean Society for the Sociology of Sport, 21*(2), 381–400.

58. Newham Monitoring Project (NMP) (2013). *Monitoring the Security Games: Final report* https://netpol.files. wordpress. com/2013/12/monitoring-the-security-games-final- report.pdf

59. Newman, H. K. (1999). Neighborhood impacts of Atlanta's Olympic Games. *Community Development Journal, 34*(2), 151–159.

60. Olds, K. (1998). Urban mega-events, evictions and housing rights: The Canadian case.

61. *Current Issues in Tourism, 1*(2), 2–46.

62. Owen, J. (2005). Estimating the cost and benefit of hosting Olympic Games. *The Industrial Geographer, 1*, 1–18.

63. Paris, C., & Blackaby, B. (1979). *Not much improvement. Urban renewal policy in Birmingham.* London: Heinemann.

64. Pillay, U., & Bass, O. (2008). Mega-events as a response to poverty reduction: The 2010 FIFA World Cup and its urban development implications. *Urban Forum, 19*(3), 329–346.

65. Porter, L., Jaconelli, M., Cheyne, J., Eby, D., & Wagenaar, H. (2009). Planning displacement: The real legacy of major sporting events/'Just a person in a wee flat': Being displaced by the CWGs in Glasgow's East End/Olympian master planning in London/closing ceremonies: How law, policy and the Winter Olympics are displacing an inconveniently located low income community in Vancouver/recovering public ethos: Critical analysis for policy and planning'. *Planning Theory & Practice, 10*(3), 395–418.

66. Pound, R. W. (2003). *Olympic Games Study Commission: Report to the 115th IOC session,Prague, July.* Lausanne: International Olympic Committee.

67. Preuss, H. (2004). *The economics of staging the Olympics: A comparison of the games 1972–2008.* Cheltenham: Edward Elgar.

68. Raco, M. (2014). Delivering flagship projects in an era of regulatory capitalism: State-led privatization and the London Olympics 2012. *International Journal of Urban and Regional Research, 38*(1), 176–197.

69. Raco, M., & Tunney, E. (2010). Visibilities and invisibilities in urban development: Small business communities and the London Olympics 2012. *Urban Studies, 47*(10), 2069–2091.

70. Rich, D. C. (1981). The Glasgow eastern area renewal scheme. *Geographical Review, 71*(2), 214–218.

71. Sadd, D. (2010). What is event-led regeneration? Are we confusing terminology or will London 2012 be the first games to truly benefit the existing local population? *Event Management, 13*, 265–275.

72. Samatas, M. (2011). Surveilling the 2004 Athens Olympics in the aftermath of 9/11: Interna- tional pressure and domestic implications. In C. J. Bennett, & K. Haggerty (Eds.), *Security games: Surveillance and control at mega-events oxon* (pp. 55–71). Routledge.

73. Sant, S. L., Carey, K. M., & Mason, D. S. (2013). Media framing and the representation of marginalised groups: case studies from two major sporting events. *Qualitative Research in Sport, Exercise and Health, 5*(3), 398–418.

74. Scanlon, K., Whitehead, C., William, P., & Gibb, K. (2013). *Building the rented sector in Scotland, Edinburgh: Homes for Scotland.*

75. Scottish Government (2014). An Evaluation of Legacy from the Glasgow 2014 Common- wealth Games. *Pre-Games Report. Scottish Government Social Research*. Edinburgh: Scottish Government Communities Analytical Services Division (CASD) and the Games Legacy Working Group.

76. Shin, H. B., & Li, B. (2013). Whose games? The costs of being "Olympic citizens" in Beijing. Environment and Urbanization, 25(2), 559–576.

77. Smith, A. (2012). *Events and urban regeneration: The strategic use of events to revitalise cities.* London: Routledge.

78. Smith, N. (1996). The new urban frontier. *Gentrification and the Revanchist city London*. Routledge.

79. Smith, N., & Derkson, J. (2002). Urban regeneration: Gentrification. In R. Shier (Ed.), *Stan Douglas: Every building on 100 West Hastings Vancouver: Contemporary Art Gallery* (pp. 62–95). Arsenal Pulp Press.

80. Smith, C. J., & Himmelfarb, K. M. G. (2007). Restructuring Beijing's social space: Observa- tions on the Olympic games in 2008. *Eurasian Geography and Economics, 48*(5), 543–554.

81. Somerville, P. (2011). *Understanding community: Politics, policy and practice. Bristol: Policy Press.*

82. Souliotis, N., Sayas, J., & Maloutas, T. (2014). Megaprojects, neoliberalisation and state capacities: Assessing the medium-term impact of the 2004 Olympic Games on Athenian urban policies. *Environment and Planning C: Government and Policy, 32,* 731–745.

83. Surborg, B., Van Wynsberghe, R., & Wyly, E. (2008). Mapping the Olympic growth machine. *City, 12*(3), 341–355.

84. Thompson, C., Lewis, D., Greenhalgh, T., Taylor, S., & Cummins, S. (2013). A health and social legacy for East London: Narratives of 'problem' and 'solution' around London 2012. Sociological Research Online, 18(2) (ISSN 1360-7804).

85. Thornley, A. (2012). "The London 2012 Olympics: What legacy?". *Journal of Policy Research in Tourism, Leisure and Events, 4*(2), 206–210.

86. VanWynsberghe, R., Surborg, B., & Wyly, E. (2013). When the Games come to town: Neoliberalism, mega-events and social inclusion in Vancouver. *International Journal of Urban and Regional Research*, *37*(6), 2074–2093.

87. Varley, A. (1992). Barcelona's Olympic facelift. *Geographical Magazine*, *64*(7), 20–24. Vigor, A., Mean, M., & Tims, C. (2004). *After the gold rush: A sustainable Olympics for London.* London: IPPR.

88. Wacquant, L. (2007a). Territorial stigmatization in the age of advanced marginality. *Thesis Eleven*, *91*(1), 66–77.

89. Wacquant, L. (2007b). French working-class banlieues and Black American Ghetto: From conflation to comparison. *Qui Parle*, *16*(2), 5–38.

90. Watt, P. (2013). *It's not for use city: Analysis of urban trends, culture, theory, policy, action 17(1). (pp. 99–118), 99–118.*

91. Zukin, S. (1995). *The cultures of cities.* Oxford: Blackwell.

# CHAPTER 7

# Marketing and Media Communications Targeted to Children as Consumers

**Blandína Šramováa***

a Comenius University in Bratislava, Faculty of Education, Raþianska 59, 813 34 Bratislava, Slovak Republic

## ABSTRACT

Marketing and media communication targeted to children as consumers is recently a topic of discussion in order to increase children consumer protection. Considering that children and youth spend a lot of time in media environment (Multi-tasking process), it is important to map the environment, including both the traditional marketing communication ways (TV, radio, DVD, video-games) and the new methods that are increasingly being preferred by children and youth (e.g. product placement, viral marketing, guerilla marketing, internet and social network environment, interactive games and quizzes, commercial website entertainment). Children are referred to as primary target group in the given communication, that is, the advertised products are addressed to they, as well as secondary target group, that is, the products are addressed to other members of the family, but the message focuses upon children as they significantly influence the shopping process in the family. This marketing communication mix, which consists of advertising, sales promotion, personal selling, public relations, uses both traditional as well as new tools and techniques which target the children's segment. In this paper, we focus on the traditional marketing techniques applied by the media environment (e.g. repetition, celebrity endorsements, and premiums) as well as new techniques emphasizing their application in educational practice.

## KEYWORDS

children; marketing and media communications; advertisement; consumerism; consumer socialization

## 1. INTRODUCTION

The marketing and media communicators' interest in child consumers has been increasing lately. There are more reasons for this particular interest, however, we can consider the awareness of children's power of persuasion on their parents' buying behavior as a crucial one. Children and the youth, not being a strong and economically powerful group, are attractive for the marketers for their ability to convince and persuade their family environment. Their power of persuasion increases with the child's age, which even the marketing and media people realize. Their communication is more directed at the child consumer. For this aim, i.e. to convince the consumer about the importance of owning a particular product or service, traditional as well as new marketing communications means are used in which school environments are not excluded from the communication process. Quite the contrary. Therefore this study focuses on introducing individual marketing communication tools, on the application of the individual forms and ways that lead to increasing materialism in child consumers as well as to their loyalty to commercial brands. Educators, psychologists and sociologists highlight the negative consequences outcoming from supporting consumerism (Schor, 2004; Opree et al.; in press, Johnson & Young, 2003; Šramová, 2014) and they ask for restrictive policies regarding advertising aimed at children. The importance of the knowledge lies in acceptance of prevention and intervention strategies that will take child's and the youth's development peculiarities into account.

## 2. MARKETING AND MEDIA COMMUNICATIONS AND THE CHILD CONSUMER

Marketing communications is a set of immediately operating activities that are of a mass, group and interpersonal character. Their aim is to achieve changes in knowledge, attitudes, behavior of a whole group in relation to a particular product or service (Hradiská & Letovancová, 2010). Lately, we have been witnessing an increasing interest of marketers in child consumers, resulting in the lay and professional public, politicians as well as academics' need for enhanced protection of children and the young. On one hand, it is necessary to analyze and exactly name the means used by the marketers in their practice, and on the other hand, it is necessary to be aware of how a child's cognitive apparatus processes stimuli, i.e. what the children ability to distinguish commercial content of marketing communiqué is. The necessity arises from the fact that children are spending more of their free time surrounded by the media, mainly by TV and the Internet (Izrael, 2013; Rideout, Foehr & Roberts, 2010) and an increasing number of advertisements containing a false idea that when a person increases their wealth, they will become more beautiful, favorite, happier and more content (Šramová, 2014). A child has the tendency to believe the presented message as they cannot distinguish a commercial content which results in reducing psychological wellbeing and enhances self-centeredness and possessiveness (Fournier & Richins, 1991; Kasser, Ryan, Couchman & Sheldon, 2004). Children do not perceive their personalities as important, they rather judge on how much they own (Sirgy et al., 2012). In this article we will deal with the essential components of marketing communications such as advertising, sales promotion, personal selling, public relations and direct marketing (Kitchen, 1999) with the emphasis on their application in communication to the child consumer.

## 2.1 Advertising

Advertising i.e. presentation of products and services to the general public, aims to inform, convince and remind an advertising message to the consumer (Vysekalová, 2009; Hradiská, 1998). In practice we encounter commercial as well as social advertisements with their task to draw our attention to urgent social problems (e.g. famine, wars, hatred, etc.). Benetton, a fashion brand, has well-known social advertising campaigns in which they present such global issues as the fight against AIDS, race intolerance, war, and so on. Commercial companies realize that through the means of global and social issues they can reinforce the brand value, as well as optimize their profits. Later we will describe which means are the most frequently used to attract, to be fast remembered and to create associations in a child consumer.

## 2.2 Sales promotion

Sales promotion is a short-term, timeshare activity that utilizes stimuli for an immediate increase in products and services purchase (Tellis, 1998, Jurášková & HorĖák et al., 2012). These stimuli are for example price discounts, coupons, contests, etc. that apply strong incentive appeals. They meet with a positive response in a child consumer mainly because they attack their natural naivety, credulity, their hobby in collecting things; in teenagers also the joy of saving some financial means.

## 2.3 Personal selling

Personal selling is regarded as one of the most effective at convincing, forming an opinion, building one's preferences for the offered product or service (Kitchen, 1999). Its effectiveness lies in a personal contact, in a twoway communication between the consumer and the vendor, in which the vendor may not only manage their communication based on the feedback they receive from the customer, but they also form long-term relationships with the customers. Children base their relationships to offered products on the popularity of a specific vendor. The preference of particular brands connected with their popularity is enhanced in personal selling and tends to persist into adulthood.

## 2.4 Public relations

Public relations is building of good relationships through maintaining favorable publicity (Kitchen, 1999). Favorable publicity reinforces good public meaning, reverses unpleasant reputation, word-of-mouth stories and events. Public relations involves in the communication with the media as well as in managing campaigns of a social dimension. We have lately been witnessing the application of social-responsible marketing realized by Causerelating marketing (CRM), social marketing, as well as fundraising. Cause-relating marketing utilizes corporate philanthropy, sponsorship, various social-oriented events support, or strategic partnerships between commercial and non-profit organizations (Marconi, 2002). Social marketing is an effort to push forward certain socially desirable attitudes, values, ideas with the aim to improve the lifestyle of people and the society (Kotler & Zaltman, 1971; Weinreich, 1999) and is realized mainly by non-profit and governmental organizations. Fundraising is a professional, targeted, organizational and scheduled activity whose aim it is to search for donors, sponsors to obtain financial or other sources for the needs of non-profit organizations (Haibach, 1998). Social, environmental, ethical projects address many consumers including teenagers. This way of marketing communication is at the same time considered an appropriate form for building, enhancing ethical, esthetic and environmental perception of children and the youth (e.g. campaigns aimed at good behavior while driving, campaigns against smoking, alcohol, etc.). There is also a wide application of this within the educational environment. Not only the connection of commercial and educational environments for compiling educational materials, evaluating pupils' performance, a close link of the school with the practice are used, but also building of a good image of the school.

## 2.5 Direct marketing

Direct marketing focuses on building relations with customers (Kitchen, 1999) in a way through which it directs its advertising activities addressable onto specific segments (based on name, address, etc.), or onto non-addressable segments (e.g. leaflets, catalogues distributed to household mailboxes). Its application is enhanced by advanced IT technologies through which it is possible to segment as well as individualize consumer's wishes, needs and interests. All the above mentioned tools of the marketing mix, i.e. advertising, public relations, sales promotion, direct marketing (under specific conditions, i.e. also a personal selling when a video-camera is involved), may be utilized also within the Internet environment. The Internet is a particularly attractive medium for children and the youth as it provides interactivity and dynamics. Banners are often creatively designed so that they use animation for a product or service promotion, which becomes more attractive for children and their attention. Marketing and communication tools are more addressable, thereby the efficiency of the advert message increases, and at the same time the use of social media is coming to the forefront. Social media represent a specific Internet environment in which people communicate, create their own and share mutual contents. A large number of people are usually concentrated within social networks; they communicate and keep in touch on a daily basis, share their experience, experiences, and so on. Social network marketing is of a special significance, i.e. using social networks by companies for brand building, enhancing brand loyalty and for sharing their advertising campaigns. Marketers thus have the option to target a large audience group that is characterized by homogeneity, interactivity and by sharing contents. For marketing communication, social networks represent an option to spread advertising messages through unusual marketing forms such as buzzmarketing, viral marketing and Internet advertising. Viral marketing uses social networks to promote goods or services through spreading texts, videos, images. Existing social networks in which people share information on products or services are used for this purpose. In viral marketing, the recipient of an advertising message becomes its distributer. Their significance lies mainly in the fact that distributing recommendations for buying a proven product has a great impact (i.e. addresses a vast number of potential consumers) and is low-cost. Buzzmarketing is a viral marketing technique that focuses on making commotion, buzz about a particular product, brand, company, event, etc., while applying elements of shock, novelty, entertainment, mystery, provocations. Another unusual way how to attract consumer's attention is the application of Guerilla marketing, i.e. a product or service promotion in an unconventional and innovative way in places where they are not expected by consumers, at minimum costs (Levinson & Godin, 1994). For children and the youth particularly, unconventional forms of presenting with a hidden commercial message, which is not detectable at first sight, are highly attractive. For this type of communication the Internet environment, where the youth spends most of their leisure time, serves particularly well. According to our findings obtained through a questionnaire method on a sample of teenagers attending grammar schools and secondary schools in the Slovak Republic (N=420, average age 17,75), 100% of them accesses the Internet at home, more than 90% of them on an everyday basis. Speaking about the frequency of the Internet use it showed that examined teenagers spend an average of 4,46 hours a day and 32,05 hours a week surfing the Internet. As the teenagers themselves stated, the Internet has become a common part of their everyday life and that they cannot imagine spending their free time without it. They do not access the Internet only through PCs, but also through their mobile phones, and the Internet thus becomes a daily companion without any limits in place of connection. It also showed that without the gender affiliation, teenagers use the Internet environment especially to satisfy their social needs in the form of keeping in touch through social networks, for downloading films and music, for obtaining information on news, for purchasing music carriers, fiction, products according to their interests. Companies use the Internet environment in a high extent for influencing consumers' attitudes. Some of them offer interactive games, quizzes for building the consumer's loyalty to the brand since childhood. At their teen age, viral marketing becomes a powerful means for influencing attitudes as at this period of their lives we can witness a strong social pressure of the peers on an individual preference for a certain brand. Another, lately frequently applied marketing technique coming to the forefront of consumers' attention is Product placement. It is a form of hidden advertising in which products, logos, brand names are placed in films, TV series, videogames, online games, entertaining shows and programs, music videos, etc. This type of a product presentation, except for the already mentioned hidden advertising, has an impact on consumer's preferences for a presented product, on creating consumer's familiarity with the product (Calvert, 2008). Calvert (2008) also draws the attention to another product placement type – websites where the sponsor's logo is placed and thus brand awareness is enhanced. The danger of the Internet environment consists also of the possibility to precisely analyze children and the youth's activities in this environment and thus use almost tailor-made persuasive marketing strategies directing to impact child consumers.

## 3. CONSUMER

Socialization by children In the next part of this article, we will briefly describe mental conditions and processes utilized by marketing communications within the media environment during consumer socialization, with the emphasis on development preferences of child consumers. In studying consumer behavior of a child, numerous authors rely on Piaget's theory of cognitive development (Piaget, 1999, Šramová, 2012, Šramová, 2014). The rate of the cognitive functions development explains the child's ability to understand marketing message; their ability to understand the selling intent of advertising (Valkenburg & Cantor, 2001; Calvert, 2008; Šramová, 2014). Consumer socialization (Ward, 1974) is most often defined as a process in which a child adopts skills, knowledge, habits, competences, and attitudes that are important for their role as a consumer. To arouse an interest in a consumer, different means of a material form (e.g. strength of the stimulus, color, contrast, unconventionality, size, dynamics), emotional form (e.g. arousing love to animals), information form (e.g. providing interesting information on the offered product), value form (e.g. the product brings satisfaction of the need of security, love, social acceptance, etc.) are used for attracting attention. Commercial messages repeating are used to increase the ability to remember the message, to build awareness as well as to enhance the image of the brand. Repeating together with presenting the model behavior (i.e. a behavior presented as an accepted norm within a certain community) through one of the means used in social learning of consumer behavior. Children audience is particularly sensitive to bright colors, loud music, and rapid changes in stimulation. Early in babies and toddlers (0-2 years old) it is possible to observe their interest in sounds, effects, animation, humour, female and children voices (Lemish, 2007). Likewise, a child is able to quickly remember a presented message especially of those advertisements where personification and animation is applied. It means when you give an inanimate object and/or phenomenon some human characteristic (e.g. such as in an advertisement for Bonduelle processed vegetables, in which little carrots, corns and peas "come into life and sing"); there are speaking animals (such as Azurit, the rabbit, promoting the Azurit fabric softener), etc. Pre-school children (3-6 years old) are attracted by messages with puppets, animation in comprehensible narratives (Lemish, 2007) as well as with fantasy elements (Šramová, 2007). In sales promotion, a bonus, i.e. added value to a purchased goods, is often attached. Very often this is a toy (e.g. a character from a Disney fairy-tale in a McDonald's Happy meal), or a coupon-/card-promoting collectibles, and so on. It is rather an effective method of effecting children desires, wishes and preferences. School-age children (7-12 years old) like messages, educational programs, sci-fi, as well as stories with heroes, human relationships. Popular TV series characters become their heroes and those who recommend certain products, as well as do celebrities who together with their peers influence brand preferences (e.g. Spider-man, Hannah Montana, David Beckham, etc.). Celebrity endorsement, when famous personalities are involved in advertising campaigns to promote a product or service, address children as well as adult audiences. The most common examples of celebrity endorsement are a promotion of cosmetic products, fashion wear, alcohol and motor vehicles. Dependence on a recommendation from a well-known show business or sports person (celebrity endorsement) may be observed even more in teenagers (12+) who prefer action scenes (in advertisements, programs and shows) and who appreciate utilitarian values while shopping on-line. Playing computer games and chatting on the Internet are their most preferred leisure activities, and by this fact teenagers become an easily accessible group for viral marketing and product placement. In the research observation of a brand impact on teenagers, a high rate of their dependence on brands occurs (Šramová, Džupina & Jurášková, 2013), especially that one relating to cosmetic products, electronics and clothing. At the same time this segment of consumers shows also a higher correlation between brand features and a teenager self-efficacy, which enhances manipulation of their attitudes to themselves as well as to their surroundings.

## 4. CONCLUSION

Marketing communication uses besides the traditional media (especially television, radio broadcasting, press media), through which it comes to consumers' awareness, also the new media (the Internet, mobile phones, etc.). Through the development of IT technologies, companies come into the consumer's environment faster, more easily and often even less costly. To reach this, they apply various hidden forms such as for instance Product placement, viral marketing, guerrila marketing, animations, and interactive games. These are sophisticated procedures that increase attractiveness of presented messages in the child consumer. An active approach of an individual to a consumer behavior appears to be one of the possible protective measures, except for legislation amendments. Such approach is referred to as consumer literacy, i.e. creating a critical consumer behavior in acquiring effective attitudes and skills focused on consumer and shopping behavior (Šramová, 2014). Increasing consumer literacy is often a part of media literacy which should become a part of the current educational process at schools as well as home environments. Its aim is to bring up an individual with a critical approach to media and marketing communications in a way that is not passive, but which make them an active recipient of media and marketing messages.

## REFERENCES

1. Calvert, S. L. (2008). Children as Consumers: Advertising and Marketing. The Future of Children 18 (1), 205-234.

2. Fournier, S., & Richins, M. L. (1991). Some theoretical and popular notions concerning materialism. Journal of Social Behavior and Personality, 6, 403-414. Haibach, M. (1998). Handbuch Fundraising. Frankfurt/Main: Campus Verlag.

3. Hradiská, E. (1998). Psychológia a reklama. Bratislava: ELITA.

4. Hradiská, E., & Letovancová, E. (1998). Psychológia v marketingovej komunikácii. Bratislava:UK

5. Izrael, P. (2013). Religiousness, Values, and Parental Mediation of Children's Television Viewing in Slovakia. Journal of Children and Media, 7 (4), 507-524.

6. Johnson, M.D., & Young, B.M. (2003). Advertising history of televisual media. In E.L. Palmer, & B.M. Young (Eds.) (2nd edition), The faces of televisual media. Teaching,violence, selling to children (pp. 265-286). Mahwah, NJ: Erlbaum.

7. Jurášková, O. & HorĖák, P. et al. (2012). Slovník marektingových komunikací. Grada: Praha.

8. Kasser, T., Ryan, R. M., Couchman, C. E., & Sheldon, K. M. (2004). Materialistic values: Their causes and consequences. In T. Kasser & A. D. Kanner (Eds.), Psychology and consumer culture: The struggle for a good life in a materialistic world (pp. 11-28). Washington DC: American Psychological Association.

9. Kitchen, P.J. (1999). Marketing Communications: Principles and Practice. Thomson: Cengage Learning EMEA

10. Kotler, P., & Zaltman, G. (1971). Social marketing: An approach to planed social change. Journal of Marketing, 35 (3), 3-12.

11. Lemish, D. (2007). Children and Television: A global perspective. Malden MA: Blackwell.

12. Levinson, J.C., & Godin, S. (1994). The Guerrilla Marketing Handbook. New York: Houghton Mifflin.

13. Marconi, J. (2002). Cause marketing: build your image and bottom line through socially responsible partnership, programs and events. Dearborn: Dearborn Trade Publishing. Opree, S.J., Buijzen, M., van Reijmersdal, E.A., & Valkenburg, P.M. (in press). Children's Advertising Exposure, Advertised Product Desire, and Materialism: A Longitudinal Study. Communication Research.

14. .Piaget, J. (1999). Psychologie inteligence (F. Jiránek, Trans.). Praha: Portál. (Original work published 1961)

15. Rideout, V.J., Foehr, U.G., & Roberts, D.F. (2010). Generation M2: Media in the lives of 8-to 18-year olds. Menlo Park, CA: Henry J. Kaiser Family Foundation.

16. Schor, J. B. (2004). Born to Buy: The Commercialized Child and the New Consumer Culture. New York: Scribner.

17. Sirgy, M.J., Gurel-Atay, E., Webb, D., Cicic, M., Husic, M., Ekici, A., Johar, J.S. (2012). Linking advertising, materialism, and life satisfaction. Social Indicators Research,107, 79-101.

18. Šramová, B. (2007). OsobnosĢ v procese ontogenézy. Bratislava: Melius.

19. Šramová, B. (2012). Vplyv kognitívneho a morálneho vývinu detí a mládeže na porozumenie mediálnych obsahov. In D. Petranová, M. Solík & N. Vrabec (Eds.), Mediálna výchova a kultúrna identita – megatrendy a media 2012 (pp. 168-176). Trnava: FMK.

20. Šramová, B. (2014). Media literacy and Marketing Consumerism Focused on Children. Procedia - Social and Behavioral Science. (in press).

21. Šramová, B., Džupina, M., & Jurášková, O. (2013). Impact of value structure on brand engagement depending on degree of self-esteem of adolescents. International Journal of Innovation and Applied Studies, 3 (4), 931-940.

22. Tellis, G.J. (1998). Advertising and Sales Promotion Strategy. MA:Addison-Wesley.

23. Valkenburg, P. M., & Cantor, J. (2001). The development of a child into a consumer. Journal of Applied Developmental Psychology, 22 (1), 61- 72.

24. Vysekalová, J. et al. (2009). Psychologie reklamy. Praha: Grada.

25. Ward, S. (1974). Consumer Socialization. Journal of ConsumerResearch, 1 (2), 1-14.

**26.** Weinreich, N. K.(1999). Hands-on social marketing: a step-by-step guide. Thousand Oaks: Sage Publications.

# CHAPTER 8

# Country as a free sample: the ability of tourism events to generate repeat visits. Case study with mobile positioning data in Estonia

**Andres Kuusik[a], Kati Nilbe[b], Tanel Mehine[a], Rein Ahas[b]**

[a]University of Tartu, Faculty of Economics and Business Administration, Narva Rd. 4-A224, 51009 Tartu, Estonia

[b] University of Tartu, Institute of Geography, Vanemuise st 46, Tartu, 51014, Estonia

## ABSTRACT

This study explores the long-term effect of events in destination marketing: the ability of events to generate repeat visits to a particular area. From this point of view, events can be treated as the free sampling method in destination marketing. When a person visits an event, they also 'get a taste' of the area in which it is taking place. And if this experience is positive, the probability of them returning later specifically to explore the area should rise. The aim of this study is to quantify the longterm effect of events in destination marketing. We analysed the post-event behaviour of more than 56,000 visitors of almost 143 small and medium-sized sports and cultural events that took place in Estonia during the years 2006-2009. We applied passive mobile positioning data in this study:. This is data, which is automatically recorded in the memory files of mobile operators. Mobile positioning enabled us to study visitors' behaviour in greater detail, both temporally and spatially. The main theoretical contributions of this study are: (1) that events are able to generate repeat visits and therefore have a long-term effect in the context of destination marketing; and (2) that this ability is dependent on the type and regularity of events. For destination managers and governments this study provides a number of suggestions on how to make more effective use of events in destination marketing.

## KEYWORDS

Destination marketing; free sampling; sales promotion; repeat visitation; Estonia; passive mobile positioning; events.

## 1. INTRODUCTION

Free sampling is a method of sales promotion frequently used in retailing. They usually make customers feel that now is the right time to purchase something (see for example Abraham, M. M. & Lodish, L. M. 1993; Ailawadi, K. L., Harlam, B. A., César, J. & Trounce, D. 2007; Bandyopadhyay, S. 2009). Free sampling in different forms is the only sales promotion method in which also a long-term effect has been detected (see for example Gedenk, K. & Neslin, S. A. 1999; Bawa, K. & Shoemaker, R. 2004; Villas-Boas, J. M. 2004; Seetharaman, P. B 2004). If we follow the principle that sales promotions make customers feel that now is the best time to buy, then in the context of destination marketing events can be treated as sales promotion activities. An event of interest – a concert for example – could be the decisive factor, which leads people to choose to visit an area at a certain time. In literature, several authors have asserted that events are used to draw in new visitors (Getz, D. 2008); (McCartney, G. J. 2005), or in the case of regular events to generate repeat visits (Kaplanidou, K. & Vogt, C. A. 2007).

Although the influence of events has been fairly extensively studied in literature, there is one aspect to which a great deal of attention has not been paid: the ability of events to generate repeat visits to a particular area. The latter should not be underestimated, as repeat visitors spend more on shopping, local transport, hotel accommodation and meals outside their hotels than first-time visitors (Wang, D. 2004) and also generate positive word-of-mouth (Oppermann, M. 2000); (Buttle, F. 2004); (Petrick, J. F., Morais, D. D. & Norman, W. C. 2001).

As such, the aim of this study is to quantify the long-term effect of events in destination marketing. We look at 143 small and medium-sized sports and cultural events (concerts, festivals and competitions) that took place in Estonia and focus on two main research questions:

- Do events have the ability to generate repeat visits to the country in which they take place?
- What kinds of features of events promote repeat visits?

We applied a new type of data in this study: passive mobile positioning (Ahas, R. & Mark, Ü. 2005); (Ahas, R., Aasa A., Roose A., Mark Ü. & Silm S. 2008). This is data, which is automatically recorded in the memory files of mobile operators. Mobile positioning enabled us to study visitor behaviour in greater detail, both temporally and spatially. We were able to identify visitors to specific areas and events at specific times, and we also used a methodology for distinguishing repeat visits.

In the first part of the study we construct the conceptual framework based on literature about free sampling and its ability to generate repeat visitation and about the role of events in destination marketing. After this we provide a brief introduction to passive mobile positioning data and the empirical study. Finally, we present our empirical findings and discuss the practical value of our results, its limitations and future avenues of research.

## 2. THEORETICAL BACKGROUND

Sales promotions are usually targeted at consumers and are designed to take immediate action to excite consumers' product interest, trial or purchase (Robinson, W. A. & Carmack, L. L. 1997); (Gilbert, D. & Clifford, J., N. 2002); (Hawkes, C. 2009). Studies have shown that sales promotions have an impact on consumers' brand choice, purchase quantity and purchase time (Blattberg, R., C. & Neslin, S. A. 1989); (Gupta, S. 1988); (Nagar, K. 2009); (Jones, S. C. & Smith, K. M. 2011). Mainly used in retail marketing but also suited to the tourism sector (Cowell, D. W. 1984); (Esu, B. B. & Ebitu, E. 2010), sales promotions have some significant advantages, including the flexibility of using several combined methods and a rapid influence on the decision-making process in regard to consumers' purchases (Hartley, S. W. & Cross J. 1988), (Mittal, M. & Sethi, P. 2011). Although such devices as discounts, coupons, prizes and collecting promotions generate rapid results, the disadvantage here is that the promotional effect is short-lived (Hawkes, C. 2009); (Hartley, S. W. & Cross J. 1988); (Mittal, M. & Sethi, P. 2011); (Bhasin, A., Dickinson, R.,

Hauri, C. G. & Robinson, W. A. 1989). In contrast, free samples as a method of sales promotion differ from other devices because they have a proven long-term effect on sales and an ability to build customer brand loyalty (Bawa, K. & Shoemaker, R. 2004); (Villas-Boas, J. M. 2004); (Seetharaman, P. B 2004). This is a result of learning (Gedenk, K. & Neslin, S. A. 1999). Free samples are one way of offering a taste of a product's qualities. If these qualities are better than those of other products, customers will prefer the new brand over others. (Villas-Boas, J. M. 2004); (Heilman, C., Lakishyk, K. & Radas, S. 2011) Also, customers prefer familiar brands to other unfamiliar brands (structural state dependence) (Seetharaman, P. B 2004). In the context of destination marketing it can also be claimed that visitors of events have the opportunity to experience the qualities of a free sample – i.e. of the country in which the event is taking place – and on this basis to form their opinions on the worth of the country. If visitors later revisit the country, it suggests that they either prefer it to others because of its better qualities or that they have made a 'safe' decision, preferring a country of which they have at least some experience over a totally unknown country.

In official tourism literature, the destination is defined as a physical space in which a tourist spends at least one night. This includes tourism products such as support services, attractions and tourist resources within a day's return travel. Destinations can be on any scale – from an entire country, region or island to a city, town or village, or a self-contained centre therein (United Nations World Tourism Organization, 2007). In destination marketing literature, 'destination' is treated as a somewhat broader concept. Buhalis (2000) posits that a destination is a combination of all locally offered products, services and experiences. Saraniemi and Kylanen (2011) have shown that destinations can be viewed as anything – a geographical area, a commodity, a unit of action, a socio-culturally constructed space, a set of institutions and actions and so on. Framke (2002) concludes that a destination is a place with an identity generated by the activities, interests, infrastructure and attractions related to it. He also claims that a destination can be anything that exists in a certain place at a certain time that offers any social activity to tourists. The meaning of 'destination' depends on the purpose of the visitor. Thus, it is possible to conclude that a destination can be:

- a place, country or geographical area, e.g. Berlin;
- a natural or artificial attraction, e.g. Berlin Zoo; or
- an event or activity, e.g. a concert or sports event held in Berlin.

Events have become an inseparable part of modern society. Their meaning exceeds the notion of tourism. They are organised for the purpose of communicating with family or friends and promoting local life and identity, culture, sport, the economy, tourism etc. (see Kavoura, A. 2007 for example). Getz (2008) proposes the following classification of events: cultural celebrations; political and state events; arts and entertainment events; business and trade events; educational and scientific events; sports and other competitive events; recreational events; and private events. The motives behind organising events are mixed, with one event rarely serving one narrow purpose. For instance, tourism events targeted at foreign tourists may simultaneously be attractive to entrepreneurs seeking to turn a profit and to local people for entertainment purposes. Similarly, local community events may be attractive to foreign tourists. Although the meaning of events and the motives of organisers have become mixed, showcasing a region and attracting tourists remains one of the most important functions of events. In this study we focus on the ability of events to generate repeat visits.

## 3. CONCEPTUAL FRAMEWORK

The conceptual framework of our study is based on the logic of free sampling and on the conclusion we made above that a destination can be a geographical region or a particular event depending on the visitor's purpose. Let us suppose that in the case of the first visit the destination is an event (see Figure 1). This means that all costs (tickets, transport, accommodation etc.) are incurred in order to 'buy' the event. Now, during the event the visitor has the opportunity to evaluate the qualities of the country in which the event is taking place, because the country is bundled together with the event as a 'free sample'. As a result, there are three possible types of behaviour:

- the visitor will not come back;
- the visitor will 'buy' the event again (arrow A in Figure 1) – in the case of regular events; or
- the visitor will come back to the country because he/she is interested in exploring it (arrow B in Figure 1). In this case the country becomes to the new destination – all costs will be incurred in 'buying' the country.

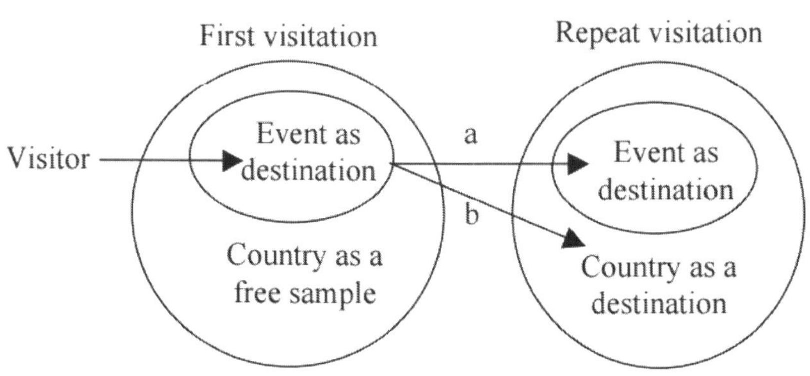


**Figure 1.** Possible destinations of repeat visitation.

In the following empirical study we measure the ability of events to generate repeat visits. In order to systemise the events and their effects, we follow the typologies established by Getz (2008).

# 4. DATA AND METHOD

Mobile positioning determines the location of mobile phones with radio waves using antennae or satellitebased geolocation. According to Shoval and Isaacson (2009), this is less costly than direct observation techniques and provides an objective snapshot of the subjects' location. Mobile positioning can be divided into passive and active positioning. Active mobile positioning is used for mobile tracking where the location of the mobile phone is determined (requested) using a special query via radio wave. Passive mobile positioning (PMP) is data that is automatically stored in the memory or log files (Call Detail Records (CDR): billing memory, hand-over between network cells, Erlang of antennae etc.) of mobile operators (Ahas, R. & Mark, Ü. 2005); (Ahas, R., Aasa A., Roose A., Mark Ü. & Silm S. 2008). The most common method of passive mobile positioning is a CDR recorded for any active phone use, also known as call activities. Any active use of a mobile phone (calls and text messages in and out, GPRS etc.) is deemed call activity. Typically, passive mobile positioning data is collected using the spatial resolution of network cells, which is referred to as Cell Global Identity (CGI).

We used data from Estonia's biggest mobile operator EMT (Estonian Mobile Telephone). EMT covers nearly 99.9% of Estonia (Estonian Mobile Telephone EMT 2010), which measures approximately 45,000 km$^2$. Market research shows that EMT has a 46% share of the local mobile phone market (TNS EMOR, 2008-09). The method of data collection and analysis was developed in Estonia in cooperation with the private company Positium LBS, local mobile operators and the Department of Geography at the University of Tartu. The database used in the study consists of a spatial and temporal register of call activities of foreign mobile phones using EMT's roaming service during the period April 2005-November 2011. The roaming service means that mobile phones registered in countries other than Estonia can be used within the Estonian network. Due to privacy issues, the database is anonymous and does not contain any back-traceable personal information about users. To recognise a person, which is essential in order to analyse repeat visits and loyalty, a randomly generated unique ID number is assigned to every phone. The ID generated by the mobile operator enables the identification of calls made by one person during the study period.

The use of mobile positioning data raises the issues of privacy and surveillance, which are fundamental ones in areas involving the monitoring or tracking of an individual's spatial and temporal movements. In Estonia, out first step was to work with mobile operators and the Estonian Data Protection Inspectorate to carefully check the accordance of data use with Estonian legislation and EU directives (Directive of the European Parliament and of the Council 95/46/EC of 24 October 1995 on the protection of individuals with regard to the processing of personal data and on the free movement of such data EU); (Directive of the European Parliament and of the Council 2002/58/EC of 12 July 2002 concerning the processing of personal data and the protection of privacy in the electronic communications sector EU). A committee consisting of external observers ruled that the data use complied with Estonian and European legislation. No personal information is connected to the data, and the generalisation level of the analysis does not allow for the identification of individuals on geographical or temporal grounds.

We used a quantitative model to pinpoint events from the time series of call activities in mobile network cells (Tiru, M., Kuusik, A., Lamp, M-L. & Ahas, R. 2010). The major criterion for pinpointing potential events was recognising a temporary increase in foreign calls by 30% compared to the three-month moving average. All 900-network cells were scanned within this framework during the study period. If such a peak was discovered in a cell, the surrounding antennae were also studied and other peaks detected. The mobile phones (i.e. anonymous IDs) that exhibited call activities during the events on the marked antenna or antennae were considered potential event visitors. All of the 'visits' of these IDs were then requested from the database for study purposes.

The second stage of the study was identifying events. For this purpose, the location and time of an event detected in the database was used. This data was compared to regional tourism calendars, local newspapers and the Internet. The filtered visit dataset combines all of the visits of those, who visited the selected events and shows how many of them had previously visited Estonia, how many were visiting for the first time and how many later returned to the country. For every visit, the time and location (antenna/cell) was recorded. An algorithm for the detection of repeat visits (Tiru, M., Kuusik, A., Lamp, M-L. & Ahas, R. 2010) was used.

We distinguished between three periods in our study (see Figure 2). The nine-month 'pre-events period' from April 2005-January 2006 was used to define first-time visitors. As we have no data about the period prior to April 2005, we defined first-time visitors as those visiting a particular event in Estonia who had not visited the country since April 2005 i.e. for at least nine months. For the purposes of the study we identified a total of 240 events, 143 of which took place from January 2006-November 2009 (the 'events period' in Figure 2) and were included in the analysis. In the case of each event in this period (marked as X in Figure 2), we pinpointed firsttime visitors and analysed their returning behaviour in the two years following the event (the dotted line A in Figure 2).

We also needed to identify events taking place from December 2009-November 2011 (the 'post-events period'). We used these events (marked as Y in Figure 2) for the analysis of visitors who visited Estonia for the first time because of events, which had taken place in 2008/2009 (see dotted line B in Figure 2). In order to make the influence of different events comparable, we only studied the two-year effect of each event.

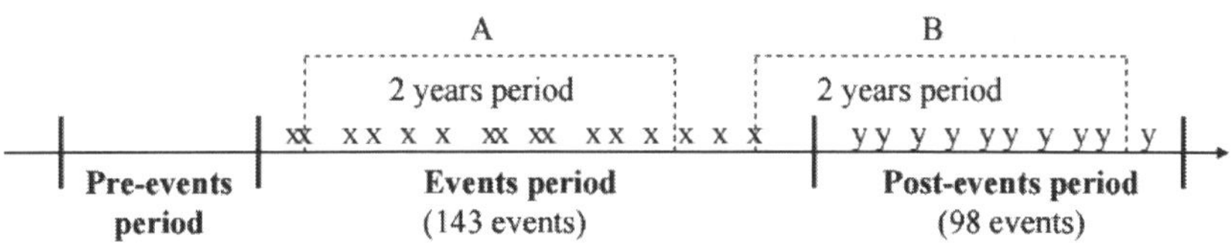


**Figure 2.** Time periods of study

An overview of the analysed 143 events, which happened during the "Events-period" (January 2006 – November 2009) is given in the Table 1. These events had 118 860 visitors, from who 56 172 were first time visitors. We divided events to subgroups based on the typology proposed by Getz (2008) with the exception that in the case of sport events we distinguished motorsport events.

We classified the events in this way on the basis of the major objective (i.e. meaning) of the event. In many cases, events are a mixture of several attractions. For example, sport events often resemble music festivals, while exhibitions can be cultural happenings. We were unable to identify political, recreational or private events. We also distinguished one-off and recurring events.

**Table 1.** Types and descriptions of identified events

| Event type | Description | Number of events | Number of visitors | Number of first-time visitors |
|---|---|---|---|---|
| Cultural | Cultural festivals | 24 | 12 883 | 5 316 |
| Arts and entertainment | Concerts and musical festivals | 34 | 68 283 | 33 671 |
| Business and trade | Fairs and trade shows | 9 | 3 726 | 1 351 |
| Scientific | Scientific conferences | 5 | 3 098 | 1 742 |
| Motorsports | Rallies and motocross | 25 | 9 036 | 3 432 |
| Other sports | All sport events other than motorsports | 46 | 21 834 | 10 600 |
| **TOTAL** | | **143** | **118 860** | **56 172** |

We tested the long-term influence of events from two viewpoints. First, we analysed the overall ability of events to generate repeat visits. We looked at how many first-time visitors came back to Estonia after the event in question; how many came back to the same event (if it was a regular event); and how many came back for an event of the same type. Secondly, we investigated whether different events wield a different long-term influence on visitors' behaviour. Here we used descriptive statistics and compared the most influential events with those which had the least influence in generating repeat visits

## 5. RESULTS

The results showed that of all 56 172 first-time visitors, a total of 15 796 (28.1%) came back to Estonia in the two years following the event in question. 1 167 (2.1%) came back to the same event, while a further 1 236 (2.2%) came back to an event of the same type. From a comparison of the 15 most influential and 15 least influential events we can see that the most influential events are regular ones and motorsports. Successful regular events that generate repeat visits include the Tallinn Rally (2006, 2008 and 2009); the Skate Marathon (2007 and 2009); the Southern Estonia Rally (2006 and 2009); and the Rakvere Dog Show (2008 and 2009). Among the events that have the least ability to generate repeat visits, no particular event type is as dominant as with events wielding long-term influence. From results we can see that in comparison with most influential events the proportions of non-regular specific sports events and regular arts and entertainment events are higher.

Compared with motorsports events, all other types generate significantly fewer repeat visits. In terms of the highest long-term impact, motorsports events are followed by cultural and other sports events. The least effective generators of repeat visits are business and trade, scientific and art and entertainment events. On average, events of different types are quite similar in encouraging visitors to come back and explore the country: the differences between average proportions are not statistically significant. Where differences do occur is in regard to visitors who come back to visit the same or same kind event. Results reveal that of all motorsport event visitors, on average more than 9% will later return to the same event, while over 7% will return to attend a similar event. Also, cultural and sport events boast rather a high proportion of first-time visitors who come back to visit the same or same type of event.

Regular events see significantly more visitors returning than non-regular events. This difference not only results from the fact that visitors cannot come back to the same event in the case of non-regular events, but that in the case of non-regular events there are also around 10% fewer first-time visitors who later return to explore the country.

# 6. DISCUSSION

Empirical analysis provided evidence that events have the ability to generate repeat visits to the country in which the events take place. We can therefore conclude that events have a long-term effect on the destination and that they can be used as part of the 'free sampling' promotional method in destination marketing. It is highly likely that visitors to an event also experience the qualities of the free sample – i.e. the country in which the event is being held – and that if they enjoy the experience, the probability that they will revisit the country (in the case of Estonia) is, on average, around 24%. This effect of a 'free sample' is based on learning as studied in marketing (Gedenk, K. & Neslin, S. A. 1999), but it can also be concluded that visitors will learn more about their 'product' from being based in the destination for several days before, during and after the event

(Framke, W. 2002). And since event tourism is primarily related to the offering and seeking of positive experiences, such tourists are predisposed to like the event and destination (Getz, D. 2008). Compared to event visitors, consumers of retail products promoted using free samples are likely more neutral or even passive towards the product (Villas-Boas, J. M. 2004); (Seetharaman, P. B 2004).

Our study also revealed that not all events are equally able to generate repeat visits, with the difference being as much as tenfold. In some cases over 70% of first-time visitors return to the country in the two years after an event, while in others only around 7% do so. Based on the 143 events studied, it can be said that an event's longterm effect in destination marketing depends on the regularity and type of the event. Here we used simple classification by field of activity such as 'cultural festivals', 'music festivals' or 'scientific conferences' as developed by Getz (2008). But this typology is problematic, as festivals and fairs can be very different. We studied repeat visits with micro data and therefore need more sophisticated features for description. For example, we divided sports events into 'motorsports' and 'other'. The comparison of 15 events with high ability and 15 events with low ability to generate destination loyalty showed that in real life much also depends on the event itself. There are some very successful and very unsuccessful events, which have extensive or, on the contrary, limited long-term effects.

In summary, the main theoretical implications of our study are as follows:

- It is possible to use the free sampling promotional method in destination marketing.
- Events have a long-term effect in the context of destination marketing because they have the ability to generate repeat visits.
- The long-term effect of events is depending on the type and regularity of the event.

Our study also presents a range of managerial implications. First, based on our study it can be seen that there is an additional good reason to use events in destination marketing. The main purpose of events in this kind of marketing is usually related to their short-term effects: attracting more new visitors and generating direct cash flow. More long-term purposes are usually related to the creation of destination image and awareness. We propose that the generation of repeat visits is also a valid purpose of events. This means that free sampling is a functional promotional method not only in retailing, but also in tourism and destination, and one that managers should start making more use of in destination marketing.

Secondly, we believe that the value of events has been underestimated by destination managers to date. The value of repeat visits generated by events tends not to be included in the value analysis of events. However, in future it could be used as a precise indicator in the break-even analysis of events.

Thirdly, and partly following on from the second point, the potential ability of a particular event to generate repeat visits can be seen as an additional criterion when destination managers are deciding which events to support and the extent to which they will support them. We can recommend to the Estonian government that regular motorsports and cultural events taking place in spring are more worth supporting. Another point is that we also need tools and measures to foster efficiency in events exhibiting a low level of performance in terms of generating repeat visits. Additionally, given that events have the ability to generate repeat visits and that even without any additional effort the rate of return among first-time visitors is, on average, around 28%, the government should consider the value of extra promotion (e.g. coupons or leaflets) among those attending the events they are supporting so as to increase the rate of repeat visits.

Finally, based on our study we can say that PMP is a very good method of identifying and analysing visitor behaviour – especially repeat behaviour. Destination managers should work with local mobile operators to obtain positioning data. This data does not include personal information and therefore is not quite as useful as a CRM system, but still provides operative feedback and a reliable analytical basis for the creation and assessment of destination marketing activities.

The limitations of this study are mainly related to the use of PMP data and to some of the assumptions we made. There are several sampling issues related to the PMP data (Ahas, R., Aasa A., Roose A., Mark Ü. & Silm S. 2008). First, the data quality is dependent on telephone use frequency, as the database is made up of call activities. Some nationalities and user groups use mobile phones more and some less actively. For example, people who make fewer calls might be underestimated or not recorded during events. Business tourists and young people use their phones more, while lower income nationalities and older people tend to use them less. Moreover, some events may be locations for heavy phone use because of conversations and emotional responses, while at other events a request to turn phones off may be issued. We also have to consider the long-term changes in the identities of tourists, as phones and phone numbers may change over time. This is something we cannot distinguish. But since the introduction of the EU Directive 2002/22/EC on universal service requirements and the right to keep the same phone number when changing operators, most users have exercised their right to keep their number.

Related to the assumptions we made, the most problematic aspect is our definition of 'first-time visitor'. We assumed that such visitors would be visitors to a particular event in Estonia who had not visited the country before – but the problem here is that we lack data about the period prior to April 2005. Therefore we do not actually know whether the first-time visitors in our study really are in Estonia for the first time or whether they had been here before April 2005. Nevertheless, we believe that this limitation had no significant impact on the findings of our study.

Based on these limitations, there are several possible avenues of future research. First, different research methods (questionnaires and interviews) or analysis methods (regression analysis) could be combined to obtain a deeper insight into the motives behind repeat visits to a country. It is also possible to pursue further comparative studies, expanding them to other countries or undertaking the same research separately for visitors from different countries. Finally, other parameters could be added to the analysis – for example, the location, timing, size or target group of an event.

## ACKNOWLEDGEMENTS

This study was carried out with the financial support of the European Social Fund, Estonia Science Foundation (Grant 7405 and 7562) and the Estonian Ministry of Education and Research (target Financing SF0180037s08 and SF0180052s07).

## REFERENCES

1. Abraham, M. M. & Lodish, L. M. (1993) An Implemented System for Improving Promotion Productivity Using Store Scanner Data. Marketing Science, Summer, Vol. 12 Issue 3, pp 248-270.

2. Ahas, R., Aasa A., Roose A., Mark Ü. & Silm S. (2008) Evaluating passive mobile positioning data for tourism surveys: An Estonian case study, Tourism Management 29(3), 469-486.

3. Ahas, R. & Mark, Ü. (2005) Location based services - new challenges for planning and public administration? Futures, 37(6): 547-561.

4. Ailawadi, K. L., Harlam, B. A., César, J. & Trounce, D. (2007) Quantifying and Improving Promotion Effectiveness at CVS. Marketing Science, Jul/Aug, Vol. 26 Issue 4, p566-575.

5. Bandyopadhyay, S. (2009) A Dynamic Model of Cross-Category Competition: Theory, Tests and Applications. Journal of Retailing 85 (4) 468–479

6. Bawa, K. & Shoemaker, R. (2004) The Effects of Free Sample Promotions on Incremental Brand Sales, Marketing Science 23(3), pp. 345– 363

7. Bhasin, A., Dickinson, R., Hauri, C. G. & Robinson, W. A. (1989) Promotion Investments that Keep Paying off. The Journal of Consumer Marketing, Vol. 6, No. 1, pp. 31-36.

8. Blattberg, R., C. & Neslin, S. A. (1989) Sales Promotion: The Long and the Short of It. Marketing Letters, Vol 1 (1), pp 81-97.

9. Buhalis, D. (2000) Marketing the competitive destination of the future. Tourism Management, 21, 97-116.

10. Buttle, F. (2004) Customer relationship management: concepts and tools. Amsterdam, Elsevier/Butterworth-Heinemann.

11. Cowell, D. W. (1984) Sales Promotions and the Marketing of Local Government Recreation and Leisure Services. European Journal of Marketing, Vol 12, No 2, pp. 114-120.

12. Directive of the European Parliament and of the Council 2002/58/EC of 12 July 2002 concerning the processing of personal data and the protection of privacy in the electronic communications sector (EU).

13. Directive of the European Parliament and of the Council 95/46/EC of 24 October 1995 on the protection of individuals with regard to the processing of personal data and on the free movement of such data (EU).

14. Estonian Mobile Telephone EMT (2010) EMT Broadcast area 2010. Retrieved 21 April 2010 from: http://www.emt.ee/web/www/liitu/miks_liitudai/ emt_plussid/leviala.

15. Esu, B. B. & Ebitu, E. (2010) Promoting an Emerging Tourism Destination, Global Journal of Management and Business Research. Vol 10, issue 1, pp 21-28

16. Framke, W. (2002) The Destination as a Concept: A Discussion of the Business-related Perspective versus the Sociocultural Approach in Tourism Theory. Scandinavian Journal of Hospitality and Tourism, 2(2), 92-108.

17. Gedenk, K. & Neslin, S. A. (1999) The role of retail promotion in determining future brand loyalty: Its effect on purchase event feedback. Journal of Retailing 75(4) pp. 433–459.

18. Getz, D. (2008) Event tourism: Definition, evolution, and reseach. Tourism Management, 29 pp 402-427

19. Gilbert, D. & Clifford, J., N. (2002) The efficacy of Sales Promotions in UK Supermarkets: a Consumer View. Journal of Advertising Research, Vol 30, No 6, pp. 315-322.

20. Gupta, S. (1988) Impact of Sales Promotion on When, What, and How Much to Buy Journal of Marketing Research, Nov, Vol. 25 Issue 4, p342-355, 14p,

21. Hartley, S. W. & Cross J. (1988) How Sales Promotion Can Work For and Against You, The Journal of Consumer Marketing, Vol 5, No 3, pp 35-42.

22. Hawkes, C. (2009) Sales Promotions and Food Consumption. Nutrition Reviews, Vol 67 (6), 333-342

23. Heilman, C., Lakishyk, K. & Radas, S. (2011) An Empirical Investigation of In-Store Sampling Promotions. British Food Journal, Vol 113, No 10, pp. 1252-1266.

24. Jones, S. C. & Smith, K. M. (2011) The effect of point of sale promotions on the alcohol purchasing behaviour of young people in metropolitan, regional and rural Australia, Journal of Youth Studies, 14:8, 885-900

25. Kaplanidou, K. & Vogt, C. A. (2007) The Interrelationship between Sport Event and Destination Image and Sport Tourists' Behaviours. Journal of Sport & Tourism, 12(3–4),183–206.

26. Kavoura, A. (2007). Advertising of national identity and tourism bureaucracy. Current Issues in Tourism, 10, 399-414.

27. McCartney, G. J. (2005) Hosting a recurring mega-event: visitor raison d'e^tre Journal of Sport Tourism 10(2), 113–128

28. Mittal, M. & Sethi, P. (2011) The Effectiveness of Sales Promotion Tools among Indian Consumers: An Empirical Study. Journal of Promotion Management. Vol 17, pp. 165-182.

29. Nagar, K. (2009) Evaluating the Effect of Consumer Sales Promotions on Brand Loyal and Brand Switching segments. Business Source Complete, Vol. 13, No. 4, pp. 35-48.

30. Oppermann, M. (2000) Tourism Destination Loyalty. Journal of Travel Research, 39, pp 78-84.

31. Petrick, J. F., Morais, D. D. & Norman, W. C. (2001) An Examination of the Determinants of Entertainment Vacationers' Intentions to Revisit, Journal of Travel Research, 40(1), 41-48.

32. Robinson, W. A. & Carmack, L. L. (1997) Best Sales Promotions, 6th vol,, NTC Business Books, NY.

33. Saraniemi, S. & Kylanen, M. (2011) Problematizing the Concept of Tourism Destination: An Analysis of Different Theoretical Approaches. Journal of Travel Research, 50(2), 133-143.

34. Seetharaman, P. B (2004) Modeling Multiple Sources of State Dependence in Random Utility Models: A Distributed Lag Approach. Marketing Science, Spring, Vol. 23 Issue 2, p263-271

35. Shoval, N. & Isaacson, M. (2009) Tourist Mobility and Advanced Tracking Technologies. New York, Routlege. Tiru, M., Kuusik, A., Lamp, M-L. & Ahas, R. (2010) LBS in marketing and tourism management: measuring destination loyalty with mobile positioning data. Journal of Location Based Services, 4(2): 120-140.

36. TNS EMOR, (2008-09) Marketing Research, Omnibus questionnaires, Unpublished manuscript.

37. United Nations World Tourism Organization, (2007) A Practical Guide to Tourism Destination Management Spain, UNWTO.

38. Villas-Boas, J. M. (2004) Consumer Learning, Brand Loyalty, and Competition Marketing Science, Winter 23(1), pp. 134–145,

39. Wang, D. (2004). Tourist Behaviour and Repeat Visitation to Hong Kong. Tourism Geographies, 6(1), 99-118.

# CHAPTER 9

# Customer Relationships through Sales Forces and Marketing Events

**Anca Francisca Cruceru[a]*, Daniel Moise[b]**

[a]The Bucharest Academy of Economic Studies, Romana Square no 6, Bucharest, 010374, Romania

[b]The Bucharest Academy of Economic Studies, Romana Square no 6, Bucharest, 010374, Romania

## ABSTRACT

Competitive organizations are always concerned about the development of management skills to enable them achieve competitive advantage in the reference market. Marketers are aware of difficulty of product development and implementation of strategies focused on consumer satisfaction. In this respect, competition is very strong; therefore, there must be taken appropriate strategies and marketing tactics that will lead to achieving these objectives. Building customer relationship marketing is considered a very important goal that can turn into a genuine organizational competence. This article aims to show the role played by sales forces, on the one hand, and marketing events, on the other hand, building a solid relationship, in the long term, with the customers. As components of integrated marketing communication, both sales forces and marketing events should be developed and supported by effective strategies and marketing tactics able to determine the competitive position of the organization. To capitalize on the benefits of sales forces, their position ought to be known in relation to marketing activities within the organization. It is also a "must" that advantages created by using marketing events should be acquainted by managers. But some of the most visible benefits refer to direct contacts directly with consumers through sales forces and create special moments through marketing events. Based on the literature in marketing at presenting the advantages orientation, these two communication techniques for developing and maintaining customer loyalty constitute a strong case for developing directions of action focused on these concepts. For this purpose, the article proposes several management measures in the use of sales forces and marketing events to build customer relationships.

## 1. INTRODUCTION

Managers of proficient companies often ask themselves questions regarding the need for application and development of certain specific methods and tools for developing long-term relationships with customers. Building relationships with customers represents a real challenge that requires knowledge and understanding of marketing concepts as well as their practical use. Starting from relationship marketing, several authors aimed to capture the essence of this concept. Thus, Weitz and Bradford (1999) consider that relationship marketing is applied to a number of different marketing activities ranging from consumer frequency marketing programs to selling activities directed toward building partnerships with key business-to-business customers. Grönroos (1994) defines relationship marketing by organization efforts directed to establishing, maintaining and developing relationships with customers and other partners, a relationship that can be divided into two parts: to attract the customer and to build the relation with the customer, so that the business objectives are accomplished. Regarding the period of time in which the relationship between the two partners develops, the organization and consumer indicates that, in terms of time, relationship marketing is focused on the long run.

Berry and Parasuraman (1991) considers relationship marketing as an approach "that concerns attracting, developing, and retaining customer relationships". The aim of the organization by implementing Relationship Marketing is to develop and maintain long-term relationships with its customers and, thus, gains a competitive advantage, which is difficult to imitate by other competitors within the market. Certainly, other objectives can be taken into account by addressing RM: customer satisfaction, share of customer, customer retention, customer loyalty etc.

Considered as a new marketing paradigm, the importance of knowledge and practical application of RM was supported by other specialists in this field, Hunt and Morgan (1994), Parvatiyar and Sheth (1994), who explained the need to focus on marketing activity, predetermining, developing and maintaining long term customer relationships.

Relationship marketing is "the ongoing process of engaging in cooperative and collaborative activities and programs with immediate and end-user customers to create or enhance mutual economic value at reduced cost", as described by the authors Sheth and Parvatiyar (2000).

Zeithaml and Bitner (2000) consider RM as a philosophy, a way of doing business, a strategic orientation that focuses on keeping and improving current customers rather than winning new ones.

Grönroos, in The Marketing Strategy Continuum, captures various implications regarding the existing differences between transaction marketing and relationship marketing, from the organization's perspective.

Therefore, the need for developing customer relationship can be sustained taking into consideration at least two reasons:

(1). Accomplishing their own marketing objectives focused on achieving a competitive position on the reference market.

(2). Attracting and keeping profitable customers for the organization.

The development of these relationships must be based on several significant attributes:

(1). Trust, both the client and seller trust each other, that trust is generated either by a previous mutually beneficial relationship, either, in case of the buyer, the competence of the seller

(2). Commitment, the sales force commitment towards the mission and values of the company, towards the sales targets requires total involvement in customer relationship.

(3). Open communication, the seller must know the buyer profile and be able to identify clear elements of description of the customer. By direct contact and by creating a bidirectional communication, the seller can learn about the consumer's needs and wants and, then, can develop sales techniques that will lead to the development of customer-salesperson partnership.

**Table 1.** Implications regarding marketing transaction and relationship marketing

| The strategy continuum | Transaction marketing | Relationship marketing |
|---|---|---|
| Time perspective | Short-term focus | Long-term focus |
| Dominating marketing function | Marketing mix | Interactive marketing |
| Price elasticity | Customers tend to be more sensitive to price | Customers tend to be less sensitive to price |
| Measurement of customer implication | Monitoring market share – indirect approach | Managing the customer base – direct approach |
| Customer information system | Ad hoc customer satisfaction surveys | Real-time customer feedback system |

Source: Adaptation from C. Grönroos, *The Marketing Strategy Continuum: A Marketing Concept for the 1990s*, Management Decision, Vol. 29 No. 1, 1991, pp. 7-13

## 2. ORGANIZATIONAL COMPETENCE

One of the objectives of the organization is to strongly compete strongly within the target market and capitalize on the competitive advantage held. However, choosing the most effective competitive strategies remains a challenge for managers. The organizational competences identification represents a priority, due to the fact that, in this manner, the organization will know what path to follow and the type of strategy to be adopted. Over time, many authors have tried to present the elements of description of these competences. Hamel and Prahalad (1994) mention the fact that these core competencies take into account a set of skills and technologies owned by the organization, which enables it to offer customers a real benefit. The authors Thompson and Richardson (1996) grouped into three distinct competencies of the organization: strategic competencies, current strategic skills policy change and learning skills, specifying that all of these skills are built based on a strategic vision of the organization indicating that it has been implemented on marketing strategy. The authors Spanos and Lioukas (2001) classify the organizational competences in three distinct categories: managerial competences, marketing competences and technological competences. On the other hand, the authors Kotler and Keller (2009) present the three specific characteristics of the core competences, namely:

(1) Sources of competitive advantage, which contributes to the perceived benefits offered by consumers,

(2) Difficult to imitate by competitors,

(3) Practical applicability for different markets.

Considering that a part of the marketing skills required inside an organization to strengthen its market position is represented by the integrated marketing communication strategy, then the communication steps initiated by organizations are very important in this respect. The development of certain complementary communication steps constitutes a solution for gaining and maintaining the competitive position as well as for attracting and retaining certain category of consumers.

## 3. SALES FORCE AND MARKETING EVENTS

Orientation towards sales force and marketing events represents an effective way to achieve these marketing objectives, due to the fact that the sales force has always played a key role in developing long – term customer relationships, and marketing events provides a direct, interactive, personal method to build relationships with these customers.

Through marketing events, the sales force can achieve a number of specific objectives, such as: attracting new categories of consumers to the promoted products/services, interaction with the actual clients of the organization, knowing the profile of these clients, and so on. The role played by the sales force differs depending on the specific stage of development and implementation of the concept within the organization. These different roles are listed in Table 2, presented below.

Therefore, due to the particularity of the activity carried out by the selling staff, the direct contact with customers and building a communicative message adapted to the profile of these clients remains one of the most effective means of communication. (Spiro and Weitz, 1990)

**Table 2.** The role played/assumed by the sales force

| Role | Sales | Marketing | Partnering |
|---|---|---|---|
| Sales force objective | Making sales | Satisfying customer needs | Building relationships |
| Sales force orientation | Short-term seller needs | Short-term customer and buyer needs | Long-term customer and seller needs |
| Critical tasks of salespeople | Convincing buyers to buy products | Matching available offerings to buyer needs | Creating new alternatives by matching buyer needs with seller capabilities |
| Activities of salespeople | Influencing customers using a hard-sell approach | Influencing customers by practising adaptive selling | Building and maintaining customer relationships<br>Organizing and leading a sales team<br>Managing conflict |
| Role of salespeople | Persuader | Problem solver | Value creator |

Source: Adapted from Wotruba (1991) and Weitz, Castleberry, and Tanner (1998) cited in Barton A. Weitz and Kevin D. Bradford (1999), Personal Selling and Sales Management: A Relationship Marketing Perspective,*Journal of the Academy of Marketing Science*, Vol. 27, No. 2, pp. 241-254

An effective model for implementation of an integrated communication step can be focused on the two communication techniques, sales force and marketing events, having as marketing objectives establishing a direct channel of communication with the target audience, attracting potential customers to the company's products, informing/updating customers about the characteristics of the promoted products, etc. In this case, in order to gain a competitive position, the selected communication techniques will be represented by organizing a themed marketing event in which the sales force will present and realize demonstrations for the advertised products.

The action plan developed for this integrated communication step includes several stages:

(1). Preparing the themed event. Choosing the theme of the event will be realized taking into account a number of factors such as: the profile of target customers, similar actions initiated or not by competitors, the nature of promoted products, the company's involvement in social responsibility projects, and so on.

(2). Defining the problem identified at the level of the selected target group. This group of consumers participating in the event will be representative for the targeted consumer segment of the company. Identifying a precise problem at the level of the target group will determine the direction towards choosing several specific actions from the created event.

(3). Situation analysis. Identifying the cause of the problem, that has a direct and powerful connection with the typology of the target group (age, sex, income, purchasing behavior and consumption, attitudes towards certain products, beliefs).

(4). Establishing communication objectives may take into account the information and awareness of the existence of a problem related to, for example, the lifestyle and healthiness of the target audience, determining the participants to the event to buy the products designed for solving the problem identified.

(5). Establishing the communication strategy should consider choosing a unique communicational message built on rational and emotional elements that have a powerful impact over the target group selected for the event.

(6). Presentation and demonstration made by the sales force within the themed event.

The sales force will be supported in this endeavor by auxiliary elements provided by the marketing department of the company. For instance, the flyers and samples given to the target group participating in the event will sustain the sales force argument. These support activities will be included in the action plan of the promotional campaign.

Specific to the activities carried out by the sales force, the presentation and demonstration will be allocated to a sales strategy based on the concept of selling in the bilateral form. (Miller and Heiman, 1989) This strategy involves, as a task, applying the concept of selling focused on identifying the existing needs at the target group level and describing the appropriate product to these identified needs. Likewise, the sales force will be responsible, at this stage of the event, for the application of the concept of client based on the presentation of the best solutions for the identified problems of the target audience. As a means of accomplishment will be used a two-way communication based on attracting the present audience at the event to the presentation made by the sales force.

(7). The end of the themed seminar is a stage in which it will be sought out the establishment of contacts between the sales force and participants to the event, as well as the actual sale of the advertised products.

(8). Results evaluation of the themed event and gathering feedback from participants implies initiating certain actions of analysis and control, in order to obtain the necessary information for continuing the integrated communication steps supported by the company.

# 4. CONCLUSIONS

It is obvious that a powerful organization focuses on identifying those organizational competences, which enable the holding of the competitive position in the field of activity. The direction towards applying an integrated communication step focused on the two components, sales force and event marketing, can substantially contribute to achieving competitive success and gaining loyal clients for the company. However, the complexity of the implementation of those communication techniques implies genuine marketing skills developed in this respect. On the other hand, we cannot talk about the efficiency of the employment of a combined event with the sales force, if the latter is not actually sustained by precise marketing activities. Attracting the target customers to the organized events and persuading them through the activities performed by the sales force constitute specific marketing objectives. By combining these techniques and developing certain thematic events suitable to the target consumer profile, there can be developed long-term customer relationships. The attractiveness of the two techniques is displayed among the public due to the interest in themed events, on one hand, and the direct involvement of sales force, on the other hand. If managers will implement a customer loyalty strategy using these communication techniques, there will be respected the correct and efficient implementation rules in order to obtain positive results. Close monitoring of all the specific stages of development of the scheduled event and clear description of sustained activities by the sales force constitute compulsory conditions that have to be respected.

The orientation towards the integrated marketing communication strategies enables the organization to achieve its communicational objectives through differentiation from the actions of other competitors. Therefore, through marketing activities oriented towards communication, preparation of sales force in order to be able to sustain presentations and demonstrations at certain marketing events may constitute a priority in the management of sales force activities.

## REFERENCES

1. Berry, L.L., & A. Parasuraman. (1991) Marketing Services. Competing through Quality. FreePress/Lexington Books, MA.
2. Grönroos, Christian. (1994). From Marketing Mix to Relationship Marketing: Towards a Paradigm Shift in Marketing. Management Decision, Vol. 32, No. 2, pp. 4-20
3. Grönroos, Christian. (1991). The Marketing Strategy Continuun: A Marketing Concept for the 1990s. Management Decision, Vol. 29, No. 1, pp. 7-13.
4. Hamel, G., &Prahalad, C.K. (1994).Competing for the Future. Harvard Business School Press, Boston, MA.
5. Hunt, Shelby, &Robert Morgan (1994). Relationship Marketing in the Era of Network Competition. Marketing Management 3(1):19-28.
6. Kotler, P., &Keller, K.L. (2009), Marketing Management, 13th edition, Pearson Prentice Hall.

7. Parvatiyar, Atul, & Jagdish Sheth (1994). Paradigm Shift in Marketing Theory and Approach: The Emergence of Relationship Marketing. In Relationship Marketing: Theory, Methods, and Application. Eds. Jagdish Sheth and Atul Parvatiyar. Atlanta, GA: Emory University, Center for Relationship Marketing, 23-30.

8. Sheth, J. N., &Parvatiyar, A. (Eds) (2000). Handbook of Relationship Marketing, Sage Publications, Thousand Oaks, CA.

9. Spanos, Y.E., & Lioukas, S. (2001). An examination into the causal logic of rent generation: contrasting Porter's competitive strategy framework and the resource-based perspective. Strategic Management Journal, Vol. 22, No. 10, pp. 907-34.

10. Spiro, R.S., &Weitz, B. A. 91990). Adaptive selling: conceptualization, measurement, and nomological validity. Journal of Marketing Research, 27(1), 61-69.

11. Thompson, J., &Richardson, B. (1996). Strategic and competitive success: towards a model of the comprehensively competent organization.Management Decision, Vol. 34, No. 2, pp.5-19.

12. Weitz, Barton A., & Bradford, Kevin D. (1999). Personal Selling and Sales Management: A Relationship Marketing Perspective, Journal of theAcademy of Marketing Science, Vol.27, No. 2, pp. 241-254.

13. Weitz, Barton, Stephan Castleberry, &John Tanner. (1998). Personal Selling: Building Relationships. 3rd ed. Burr Ridge, IL: McGraw-Hill. In Barton A. Weitz, & Kevin D. Bradford, Personal Selling and Sales Management: A Relationship Marketing Perspective, Journal of the Academy of Marketing Science, Vol.27, No. 2, pp. 241-254.

14. Wotruba, Thomas. (1991).The Evolution of Personal Selling. Journal of Personal Selling and Sales Management, 11 (Summer):

1-12.In Barton A. Weitz, & Kevin D. Bradford, Personal Selling and Sales Management: A Relationship Marketing Perspective, Journal of the Academy of Marketing Science, Vol.27, No. 2, pp. 241-254.

15. Zeithaml, V. A., & Bitner, M. J. (2000). Service Marketing, 2nd Edition, Irwin-Mcgraw Hill USA.

# CHAPTER 10

# Strategies for interactive communication, with external and internal stakeholders, in events' marketing

**Moise Daniel[a]*, Georgescu Bogdan[b], Zgură Daniel[c]**

[a]The Bucharest Academy of Economic Studies,
Mihai Eminescu Street 13-15,Bucharest, Romania

[b]The Bucharest Academy of Economic Studies,
Mihai Eminescu Street 13-15,Bucharest, Romania

[c]The Bucharest Academy of Economic Studies,
Mihai Eminescu Street 13-15,Bucharest, Romania

## ABSTRACT

During the development of the events a key element is represented by the interactive communication, realized with all the persons implicated in the event; the company that realizes the event, the organization that ordered the event, the participants. It can be rightly stated that the success and the impact of an event largely depends also on the way in which it is communicated with the participants, even with the authorities. During the events, the interactive communication must be very good realized and managed because some information must be delivered only to some employees, other to all of them or only to those implicated. The restrictions are assigned mainly by the budget allocated to that specific event or by the regulations in force in the area where the specific event unfolds, but also the stakeholders that can be internal or external.

# 1. INTRODUCTION

The specific of the processes developed in the interactive marketing complaints the devise of some strategic options that serve as basis of some actions of operative management out of which are distinguished: the flow management ( also found as the string management), met widely in the development of some large scale events from the sport, cultural events, festivals field etc. The neglecting of the issue is reflected on multiple plans, the most frequent to the situation claiming another series of strategic options that make the object of the conflicts between segments management. The situation is frequently met in the sporting events where the conflict between "galleries" degenerates in open conflicts that seriously affect the development of the event itself. But the conflicts between segments management represents only one facet of a more general strategic option known as multiple segments management, a component specific to the interactive marketing generated by the co-existence, in the moment of the creation and delivery of the events, of more segments. The companies use a wide range of techniques through which they ensure the meeting of the requirements of each and every segment. For instance, in case of some concerts, on the same stage can perform in the same time two different bands, as genre and not only and the participants can choose which one of the two bands they want to listen to. But of this "luxury" cannot benefit all the events, usually the organizers, after the segmentation of the audience; direct the whole event by taking into consideration the segment with the highest weight. (Clow & Baack, 2004) Some authors consider that the organization communicates its image and the one of its products/ services offered since the entrance of the consumer at the place of the development of the event, together with the first contact with the personnel or simply through the atmosphere created. The effect of the communications can be positive or negative. Any company, that sends information and messages about the products/ services provided, must be preoccupied by their integration in a unitary system in order for the consumers to be able to create an overall image about the offer of the organization.

According strictly from this point of view of the proper interactive communication, the organizers of events have available the following strategic options (Goldblatt, 2005):

a. billboards, flyers), information indicators, user guides, environment;

b. the use of some "interpersonal" communication means of the type: personnel in contact commercial personnel (in case of the fairs ,exhibitions, etc. 0 client ("orally" promotion).

## 2. EXTERNAL COMMUNICATION

The companies can appeal to a large variety of events for communicating with the target audience, both from the external and internal environment of the company. If the realization of some events is optional depending on the budgets assigned to the communication, instead others are compulsory, such as the General

Assembly. In this way, can be realized a table with different strategic options which an organization can choose for realizing an event. These strategic options depend of course at first also upon the financial, human and material resources that the given organization disposes of for the realization of the specific event. The most important criteria that were taken into account, with the mention that can be added others are: the type of event, the way in which it is realized (in direct labour, outsourced or joint), the location where the event takes place ( must be taken into account if the organization has space assigned and sufficient for the development of the event because in case that the space is too small for the number of participants their experience can be unpleasant), the period in which the event takes place (it firstly must be considered also the availability of the participants to come to the event) and the participation tax ( if for some events the participation or the entrance is free, for other events there exists a participation tax of which quantum increases in direct proportion with the demand for such an event).

Essentially, the chosen strategy is described by a combination of alternatives constituted by taking into consideration each and every criterion. In this way can be chosen different strategic options (table 1) that must firstly take into account also the budget, human and material restrictions for the realization of different types of events and secondly the possible audience of the event concerning the period, tax and location in which it will take place. For each event there may be chosen different strategies, and also applied particular methods depending on the targeted segment, as an organizer might send free of charge invitations for VIP's as the other have to pay. For example, in the case of a conference (1.B), the chosen strategy could look as follows: realized in direct labour (mainly externalized 2.a), the place of development: the headquarter of the organization (in another location 3.a), the period in which the conference takes place (during the week 4.b) and the participation tax to be applied to the participants from outside the institution (free entrance usually for sponsors 5.a).

**Table 1.** The external communication through event strategies

| **Criteria:** **1. The type of event** | **2. Realized in:** **direct labour / externalized / joint** | **3. Location:** **at the headquarter of the organization / in another location** | **4. Period:** **weekend / during the week** | **5. Tax for the participants:** **with tax/ free of charge** |
|---|---|---|---|---|
| A. Sponsorship / Patron | a. mainly externalized<br>b. joint | a. in another location, usually where the sponsored event takes place | a. when the event takes place | a. free entrance usually for sponsors<br>b. entrance or participation tax |
| B. Conferences | a. mainly direct labour<br>b. externalized<br>c. joint | a. usually at the headquarter of the organization if it disposes of the necessary space<br>b. another location | a. weekend<br>b. during the week | a. participation tax<br>b. discount for particular members |
| C. Symposiums | a. externalized<br>b. direct labour c. joint | a. usually in other locations | a. during the week<br>b. weekend | a. participation tax<br>b. free for sponsors |
| D. Seminars | a. direct labour<br>b. externalized<br>c. joint | a. in another location, usually where the sponsored event takes place | a. usually during the week | a. participation tax<br>b. free of charge |
| E. The General Assembly | a. direct labour | a. at the headquarter of the organization if it disposes of the necessary space<br>b. another location such as conference rooms | a. during the week | a. free of charge |
| F. International, national and local fairs and exhibitions | a. direct labour<br>b. joint | a. at the place where the fair or exhibition takes place | a. when the event takes place | a. entrance tax<br>b. rarely free of charge |
| G. Launching of products / services / rebranding | a. direct labour<br>b. joint | a. at the headquarter of the organization<br>b. other locations such as retail outlets of the products | a. when it is decided the launching | a. free of charge<br>b. with participation tax |
| H. Touring performances | a. direct labour<br>b. joint | a. in another location | a. during the week<br>b. weekend | a. free of charge |
| I. Workshops | a. direct labour<br>b. joint | a. usually at the headquarter of the organization if it disposes of the necessary space | a. during the week<br>b. weekend | a. free of charge<br>b. with participation tax |
| | | b. another location | | |
| J. Open Day | a. direct labour | a. at the headquarter of the organization | a. decided by the organization | a. free of charge |

# 3. INTERNAL COMMUNICATION

Regarding the internal communication strategies in the events marketing have some special features. The internal communication represents a component of the marketing mix specific to the marketing from the interior of the company having an essential role in ensuring the accordance between the promised events (offered) that address to the audience from the internal environment and the ones created and delivered in the interactive marketing. As communicational system, the internal communication is realized between the personnel of a company that executes activities for the promised services and for the ones created and delivered.

Consequently, the internal communication is indissolubly related to the "personnel" of the company that communicates through the event.( Filip, 2011) Therefore, it aims at the attraction, support and agreement of the personnel, the increase of the degree of its involvement in the realization of the event. It cannot be omitted neither the fact that reaching the audience from an organization is almost as hard as reaching the audience represented by the general public. It can be remembered also the fact that it is preferable to be communicated with the employees, just before he enters in the organization, during the time in which he activates inside it, and also after the end of the contract.

In this way, the internal communication of the marketing ranks with the managerial communication, at the superior management level, having a specific role, the one of contributing to the attaining of the objectives stipulated in the marketing plans of the organization. Its objectives are derived from the ones of the marketing, being included in some specific programs, using specific means of internal communication. (Olteanu, 2005)

In the plan of action, for the development of the event, there will be taken decisions about the message and mean in which it will be delivered to the audience, taking into consideration the moment and the right place in order for this to reach the target audience. It must be considered also the period of time necessary for the realization of researches on the employees, the writing, the editing and the realization of the message in order for this to be as brief and on subject. The means of internal communication used are the following (JCI, 2009):

Newsletters in a written form, edited in an efficient form once per month; Newsletters in electronic format that attain precisely the target audience, the only condition would be that they use the digital station. In them can be inserted also hyperlinks to the web page of the organization where those interested can find out more information. It is indicated for the electronic newsletter to be sent weekly, to be concise and easy to be found information in it, in order for this to be read in the shortest time possible;

Web site in which besides the information that address to all the stakeholders can exist also private areas where only the employees of the organization can have access. This must be frequently updated because otherwise it denotes indifference upon the visitors and the users of the site. Also, the site must be accessible and easy to surf in, not to use jargon and abbreviations that only few of the users know;

Other printed materials such as brochures, internal notes, announcements posted on the bulletin board etc., having the role of announcing the employees of the organization about the actual and future events. The messages on this path must include also contacts of the person or persons that, on demand, can offer more information about the event.

The social media permits the use of forums, chats that realize the interaction in real time, the posting of an announcement or a discussion subject, to be debated by the ones interested. Can be used both networks of social sites and blogs.

The organizations that communicate with their own employees can recourse to a wide range of events through which they can send a diversity of messages for attaining the marketing objectives. (Cetină, 2009). As in the case of the of the strategies of external communication through event can be taken into consideration the same set of criteria (table nr. 2) the type of event, the mean in which it is realized, the location, the period and the payment form (if the employee can participate for payment or free of charge at the relative event). (Masterman & Wood, 2006) Some events for employees are free of charge such as teambuilding or trainings, sometimes these must sign additional fidelity contracts for the organization that had discharged the participation to such events, for a determined period and in the case in which they want to collaborate with another company being forced to pay the consideration of the training programs because the organization will not recover the performed investment. And in this situation, the strategy of the communication chosen is constituted by a combination of strategic alternatives that have as a basis each of the criteria listed above. An example of strategy can look like: the type of event- training 1.A, form of organization- externalized (2.a), location the headquarter of the organization (3.a), the period of development weekend (4.b), payment method for payment (5.a).

# 4. CONCLUSIONS

These strategies modulate and apply according to the human, material and financial resources of eachand every organization, not being able to be considered generally valid, insensitive to the type of the organization, taking into account also the particularities of its internal and external environment and stakeholders.

**Table 2.** Internal communication through event strategies

| Criteria<br>1. The type of event | 2. The form of organization can be realized in:<br>direct labour/ externalized/ joint | 3. Location: at the headquarter of the organization or in another location unless the headquarter of the organization | 4. The period:<br>weekend /<br>during the week | 5. Payment form<br>( contribution from the employee):<br>paid /<br>free of charge |
|---|---|---|---|---|
| A. Training | a. externalized<br>b. joint | a. regularly at the headquarter of the organization if it disposes of the necessary space<br>b. another location | a. during the week<br>b. sometimes in the week-end | a. paid<br>b. free of charge but with a fidelity contract |
| B. Teambuilding | a. externalized<br>b. joint | a. usually in another location<br>b. at the headquarter of the organization if it disposes of such a space | a. weekend<br>b. during the week | a. free of charge<br>b. free of charge but with a fidelity contract |
| C. Seminars / Webinars | a. direct labour<br>b. joint | a. at the headquarter of the organization<br>b. in another location | a. during the week<br>b. weekend | a. free of charge<br>b. free of charge but with a fidelity contract |
| E. Happy Hour | a. direct labour | a. in another location such as, for example: pubs, terraces , clubs | a. during the week | a. paid<br>b. free of charge |
| F. Religious Holidays | a. direct labour<br>b. joint | a. at the headquarter of the organization if it disposes of the necessary space<br>b. another location | a. during the week | a. usually paid<br>b. free of charge |
| G. Special occasions | a. direct labour<br>b. joint | a. especially at the headquarter of the organization<br>b. at the retail outlet of the products/ services | a. during the week | a. free of charge |

## ACKNOWLEDGEMENTS

This article is result of the project POSDRU/88/1.5./55287 "Doctoral Program in Economics at European Knowledge Standards (DOESEC)". This project is co-funded by the European Social Fund through The Sectorial Operational Program for Human Resources Development 2007-2013, coordinated by The Bucharest Academy of Economic Studies in partnership with West University of Timisoara.

## REFERENCES

1. Filip, A. (2011). Contribution of internal marketing to increasing employee satisfaction, improving service quality and strenghtening business relationships. Quality-access to success, Vol. 12, No. 123 (4-August), 43-46.

2. JCI Marketing Manual, United States of America, (2009); **http://www.jci.cc/downloads/en/marketingmaterials/othermaterials**

3. Olteanu, V. (2005). Management - Marketing, Publishing House Ecomar, Bucharest, 76.

4. Cetină, I. (2009). Marketingul serviciilor fundamente şi domenii de specializare, Publishing House Uranus Bucharest.

5. Masterman, G & Wood, E. (2006). *Innovative Marketing Communications, strategies for the events industry,* Publishing House Linacre Hous, Jordan Hill, Oxford.

6. Clow, K. & Baack, D. (2004). Integrated Advertising, Promotion, and Marketing Communications, 2nd edition. Publishing House Pearson Prentice Hall.

7. Goldblatt, J. (2005). *Special Events: Event leadership for a new world*, Publishing House John Wiley and Sons Inc., New York.

# Index